The Rise of Neuromorphics
A future with complex adaptive artificial brains

Contents

Preface

This book is published under a pen name. The reason is that I do not want to be contacted. Ideas are ideas even when not attached to names. This book is about an idea. I hope this idea will spark some interest among technologists to make new and exciting computer chips.

Maverick Boltz, November 2023

1 Neuromorphic systems as complex adaptive systems

A few weeks back I attended a conference on Neuromorphic Computing in Santa Fe, New Mexico. The conference was called *ICON, International Conference on Neuromorphic Systems*. This is one of the main conferences on neuromorphic computing. It is organized by Oak Ridge National Laboratory (ORNL), a major research institution located in Oak Ridge, Tennessee. It is one of the largest science and energy national laboratories within the United States Department of Energy (DOE) system. Established in 1943, initially for atomic weapons research during World War II, ORNL has evolved into a multidisciplinary research facility with a broad scope of scientific and technological investigations. If there is something that we have learned from Nolan's movie Oppenheimer is that there is a lot of science going into the making of both conventional and atomic weapons. The location of the 2023 conference should be also a call to arms for many complex science researchers. Santa Fe is a hotspot for many fancy ideas. It is located between two of the most important National Laboratories in the US: Sandia National Laboratory, near Albuquerque, and Los Alamos National Laboratory, in Los Alamos. Los Alamos is now in the news again because of the movie Oppenheimer. Santa Fe, on the other hand, is also where the Santa Fe Institute (SFI) resides, a prestigious research center that was and still is at the center stage of Complex Systems. Yet, I did not see any talk by Complex Science researchers at the ICONS conference, mostly engineers and a few physicists.

I instead discovered that some weeks earlier, in June, there had been another conference, a Symposium organized by the Santa Fe Institute, called *Collective Intelligence: Foundations + Radical Ideas*, in Santa Fe. For two months, then, Santa Fe gathered hundreds of experts in adjacent disciplines, tackling intelligence from a variety of angles.

One cannot describe what a neuromorphic system is without discussing some of the details of biological systems able to learn. In complex systems, the brain is often called a Complex Adaptive System, or CAS. A complex adaptive system is essentially something that can change and adapt in response to its surroundings. CAS emerged in the 90s from the study of emergence in complex systems. When we talk about this in the context of the brain, it means that our brains are incredibly flexible and can learn from experiences.

Think of the brain as a super complex system made up of billions of tiny parts called neurons that are all connected together through even tinier connections called synapses. This complexity is what gives our brains the power to think, remember things, and react to the world around us.

Neuromorphic computing is about trying to copy the brain's abilities using technology. It is an attempt to make computers work more like our brains. Instead of using traditional methods, neuromorphic computers use circuits and signals that are inspired by the way our brains work. This can make them much better at things like recognizing patterns and using less power to do it.

Brains are complex adaptive systems *par excellence*, but there are examples of seemingly much simpler complex adaptive systems everywhere.

When we look at ant colonies, we see a remarkable example of a complex adaptive system. Think of an ant colony as a bustling city of tiny insects, each with a specific job to do. This system is not just a random gathering of ants; it is a highly organized and adaptable community.

Ants in a colony can adjust their behavior based on changing conditions. For instance, they can change their foraging patterns if they encounter a new food source or adapt to different weather conditions. This adaptability is a key feature of a complex adaptive system.

Ants communicate using chemical signals and can relay information to the colony. If one ant finds food, it leaves a chemical trail for others to follow. This kind of information sharing and response to feedback is another hallmark of complex adaptive systems.

Ant colonies also display emergent behavior, meaning that the collective actions of individual ants can lead to surprising and organized outcomes. For example, they can build intricate underground nests, organize efficient foraging routes, and defend the colony against threats. Yet, there is no evidence that the individual ant has any knowledge of the overarching goal of the colony, but only follows a simple set of rules.

The way ant colonies work, with their decentralized decision-making and adaptability, has even inspired algorithms used in computer science and optimization problems. It is a fascinating example of how nature has evolved complex systems that can adapt and thrive in a wide range of environments, and it offers insights into how we can design better systems and technologies based on these principles.

There are other types of complex adaptive systems, for instance, slime molds. Slime molds are unique organisms that exemplify the characteristics of complex adaptive systems. They exist as neither plants nor animals but rather as a type of protist. What makes slime molds truly captivating is their exceptional ability to adapt and solve problems using unorthodox methods.

Slime molds consist of a network of single-celled organisms that come together to form a multicellular body. When faced with challenges such as locating food or navigating obstacles, slime molds showcase remarkable adaptability.

Slime molds demonstrate impressive problem-solving skills. Upon encountering a food source, they extend their network of cells toward it. During exploration, they optimize their routes to maximize nutrient intake while minimizing energy expenditure, highlighting their efficient problem-solving capabilities characteristic of complex adaptive systems.

These organisms exhibit decentralized decision-making. Slime molds efficiently distribute their cells to explore their environment and discover the most advantageous routes to resources, much like the operation of complex adaptive systems.

Slime molds display resilience in the face of adversity. They can endure harsh conditions and rebound when circumstances improve. Furthermore, they possess a form of memory and learning, enabling them to adapt more effectively based on past experiences.

Similar to ant colonies, slime molds exhibit emergent behavior. While individual cells lack a grand plan, they collectively form intricate patterns and solve complex mazes through self-organization, emphasizing their capacity for emergent behavior.

Researchers have discovered that slime molds offer valuable insights into optimization problems, transportation networks, and even urban planning. They serve as a reminder that even seemingly simple organisms can possess extraordinary adaptive capabilities, providing inspiration for solving complex problems and designing efficient systems across various domains. Slime molds represent yet another example of nature's ingenious solutions that inform and inspire innovations within the realm of human endeavors.

Collective intelligence, within the framework of complex adaptive systems, can be understood as the emergent property that arises when a group of individuals or components interact and collaborate to collectively solve problems, make decisions, or generate insights that surpass the capabilities of any single member. It reflects the idea that the whole, in terms of problem-solving and decision-making abilities, is greater than the sum of its parts. This idea of emergence was introduced by physicist Phillip Anderson in a well-known article.

In the context of complex adaptive systems, several key principles apply to collective intelligence:

1. **Decentralized Decision-Making:** Collective intelligence often relies on decentralized decision-making, where individual agents or components within the system act autonomously based on local information and rules. This decentralized approach allows for adaptability and responsiveness to changing conditions.

2. **Emergent Behavior:** Similar to other complex adaptive systems, collective intelligence exhibits emergent behavior. The interactions and contributions of individual entities lead to the emergence of collective behaviors, solutions, or insights that could not have been predicted solely by examining the behavior of isolated agents.

It is then interesting to compare these ideas to the modern literature on

Neuromorphic Computing. Neuromorphic computing systems are typically designed as specialized hardware and software configurations inspired by the human brain's neural structure and function. These systems aim to replicate the brain's computational capabilities, such as pattern recognition and adaptive learning, while often employing highly parallel processing and low-power consumption. While they share some similarities with complex adaptive systems, they are not typically considered as such due to specific design principles and objectives.

Designed for Specific Tasks: Neuromorphic computing systems are usually designed with a specific application or task in mind, such as image recognition, natural language processing, or robotics control. They are tailored to perform these tasks efficiently and may not possess the same level of adaptability and versatility as complex adaptive systems.

The limitations I see in the modern literature (at large) are briefly described as follows:

Structured and Engineered: Neuromorphic systems are engineered to replicate specific aspects of neural processing, often with a predefined architecture and connectivity. Unlike complex adaptive systems, which can exhibit emergent and unanticipated behaviors, neuromorphic systems are designed to operate within known parameters and follow predefined algorithms.

Limited Adaptation: While neuromorphic systems can learn and adapt to some extent, their adaptation is typically guided and controlled by algorithms and training data. They lack the self-organization and autonomous adaptation often associated with complex adaptive systems.

Focused on Efficiency: Neuromorphic computing emphasizes energy efficiency and real-time processing, which may lead to certain trade-offs. This focus on efficiency and speed can limit the degree of flexibility and adaptability in favor of meeting specific computational goals.

Absence of Self-Organization: Complex adaptive systems often exhibit self-organization, where system components interact and adapt without central control. In contrast, neuromorphic computing relies on precisely engineered components and algorithms, reducing the degree of spontaneous self-organization.

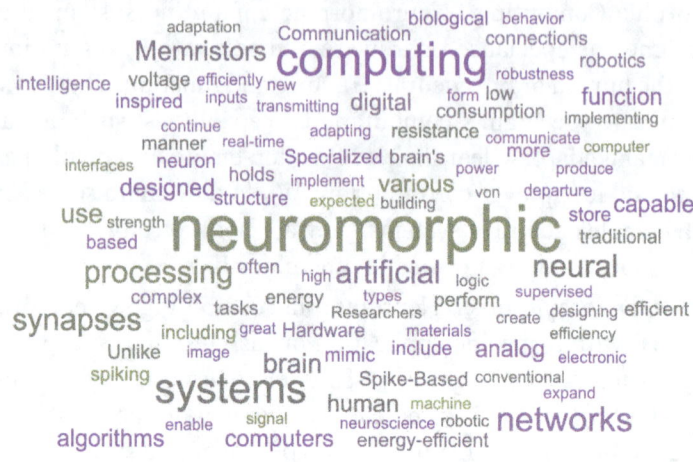

Figure 1.1: Buzzwords around neuromorphic computing.

One of the key ideas behind complex adaptive systems is the notion of *emergence*, but this is missing almost entirely in the literature.

The notion of emergence is a fascinating and fundamental concept in various fields of science and philosophy. It refers to the phenomenon where complex systems display novel properties, behaviors, or structures that cannot be explained or predicted solely by examining the individual components of the system. In other words, emergence suggests that the whole system is greater than the sum of its parts, and new properties arise from the interactions and relationships among these parts.

Emergence can be observed in diverse contexts, from physics to biology, sociology to economics. It has profound implications for our understanding of complexity and has challenged reductionist approaches that seek to explain everything by breaking it down into its smallest components.

The concept of emergence itself emerged from the literature in chaos theory, a branch of mathematics and physics that gained prominence in the mid-20th century. Chaos theory explores the behavior of dynamic systems that are highly sensitive to initial conditions. While studying chaotic systems, scientists and mathematicians began to notice patterns and behaviors that couldn't be easily predicted or explained by linear equations or traditional reductionist methods.

Chaotic systems must be nonlinear, meaning that small changes in initial conditions can lead to vastly different outcomes. These systems can exhibit sudden, dramatic shifts or bifurcations that give rise to unexpected behaviors. Chaotic systems can display complex, intricate, and seemingly random behavior over time. This complexity is not readily reducible to the interactions of individual components. In chaotic systems, patterns and structures can self-organize without external guidance. This self-organization leads to the emergence of coherent and often unpredictable behaviors. A hallmark of chaotic systems is their sensitivity to initial conditions, often referred to as the "butterfly effect." This sensitivity can lead to unpredictable emergent phenomena.

The idea of emergence became more explicit as scientists entangled with the limitations of traditional reductionist approaches to understanding complex systems. Emergence challenges the notion that everything can be explained by reducing it to its smallest parts and highlights the importance of studying systems as wholes. It underscores the idea that the interaction and interdependence of components can give rise to new, unexpected, and often beautiful properties that are not apparent when examining individual elements in isolation.

At the heart of understanding emergence in the context of the brain lies the intricate structure and functioning of neurons. Neurons are the fundamental units of the nervous system, specialized cells designed for transmitting and processing information. Their remarkable ability to communicate through electrical and chemical signals forms the basis of all brain activity.

Neurons as tiny, highly specialized messengers. Each neuron consists of a cell body, dendrites (which receive signals), an axon (which transmits signals), and synapses, the junctures through which these messages are relayed. These synapses are the crucial points of connection, where information is passed from one neuron to another. This complex network forms the basis of the brain's computational power.

The elegance of the brain emerges from the sheer number of neurons and their connections. Billions of these units work in harmony, creating a symphony of electrical activity. This interconnectedness is where the magic happens. When neurons fire, they send signals down their axons, which then influence other neurons through their synapses. This intricate dance of electrical impulses and chemical signals gives rise to the immense complexity and adaptability of the brain.

The real marvel, however, is not just in the sheer quantity of neurons, but in the quality of their connections. It is not a simple one-to-one relationship; it is a web of relationships, with each neuron potentially connected to thousands of others. This allows for an astonishing degree of parallel processing. While individual neurons may seem simple, the collective intelligence of the entire network emerges from their coordinated efforts.

Intelligence, as we understand it, arises from this emergent phenomenon. It is not confined to any specific neuron or localized region of the brain. Rather, it's a property that emerges from the collective behavior of the entire neural network. This collective intelligence enables processes such as learning, memory, decision-making, and problem-solving.

In essence, the brain's intelligence is not a top-down command but an emergent property arising from the interactions of its constituent neurons. It's a beautiful example of how complexity can emerge from simplicity, and how the whole can be far greater than the sum of its parts.

Let us now put this into practice, in a mathematical form.

The Hodgkin-Huxley model is a mathematical representation of the behavior of neurons. It describes how neurons generate and propagate electrical signals, known as action potentials. Hodgkin and Huxley obtained, in a pioneering mathematical framework, the generation and propagation of action potentials in neurons. It was conceived by Sir Alan Hodgkin and Sir Andrew Huxley in the early 1950s. This model stands as a monumental achievement in neuroscience, profoundly influencing our comprehension of the nervous system.

Prior to Hodgkin and Huxley's work, there existed a rudimentary understanding of how nerve impulses were initiated and transmitted. While it was acknowledged that nerve cells communicate through electrical signals, the precise underlying mechanisms remained elusive.

Hodgkin and Huxley embarked on a series of groundbreaking experiments, conducted primarily on the giant axon of the squid. These experiments involved inserting microelectrodes into the axon to measure the flow of ions during an action potential. The data gathered from these experiments formed the foundation upon which the Hodgkin-Huxley model was constructed.

The model itself comprises a system of differential equations that encapsulate the dynamics of ion currents (sodium, potassium, and leakage) and the activation variables (m, n, h) controlling ion channel states. Through their meticulous research, Hodgkin and Huxley formulated these equations to accurately represent the complex interplay of currents and gating variables within neurons.

This model not only provided a quantitative description of nerve impulse generation but also offered a theoretical framework for understanding a wide array of neuronal behaviors. It became a cornerstone of neuroscience, laying the groundwork for further studies on neural excitability, synaptic transmission, and the functioning of the nervous system as a whole. Hodgkin and Huxley's contributions have left an indelible mark on the field, shaping the way we perceive and study neural phenomena. If you are not mathematically inclined, you can skip what follows and go to page 12.

The model is based on a system of differential equations. We define the following quantities:

- C_m as the membrane capacitance,
- V_m as the membrane potential,
- I as the current input,
- g_{Na}, g_K, g_L as the maximum conductances for sodium, potassium, and leakage currents respectively,
- E_{Na}, E_K, E_L are called Nernst potentials, and represent for sodium, potassium, and leakage respectively,
- m, n, h as the activation variables for sodium, potassium, and leakage channels.

The Hodgkin-Huxley equations are as follows:

$$C_m \frac{dV_m}{dt} = I - g_{Na}m^3h(V_m - E_{Na}) - g_K n^4(V_m - E_K) - g_L(V_m - E_L)$$

$$\frac{dm}{dt} = \alpha_m(V_m)(1 - m) - \beta_m(V_m)m$$

$$\frac{dn}{dt} = \alpha_n(V_m)(1 - n) - \beta_n(V_m)n$$

$$\frac{dh}{dt} = \alpha_h(V_m)(1 - h) - \beta_h(V_m)h$$

where α and β are voltage-dependent rate functions.

This system of equations models the behavior of a neuron in terms of ion currents (sodium, potassium, and leakage) and the dynamics of the activation variables (m, n, h) which control the opening and closing of ion channels.

By solving these equations, you can simulate how the membrane potential of a giant squid axon changes over time in response to different inputs and initial conditions. An example of the behavior of the Hodgkin-Huxley model is shown in Fig. 1.2. You can see the spikes emerging from the application of a current in the model.

This model can be further simplified, while providing similar nonlinear behavior. This is called the integrate-and-fire model. We will see in a moment why this is important.

The integrate-and-fire model retains the focus on membrane potential dynamics, assuming that the neuron integrates incoming currents until a threshold is reached, triggering an action potential. This simplification keeps the concept of an action potential while discarding detailed ion channel dynamics.

The mathematical representation of the integrate-and-fire model is:

$$C_m \frac{dV}{dt} = I_{input} - V$$

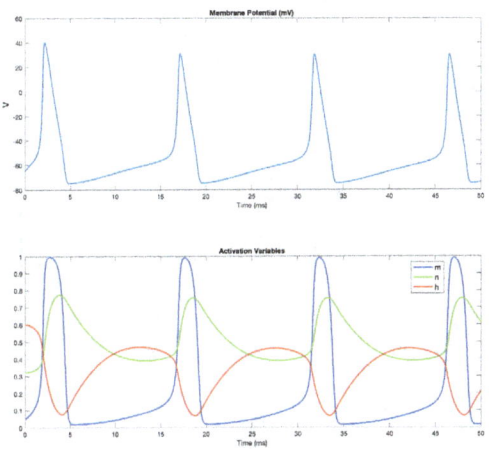

Figure 1.2: Spiking behavior of the neuron in the Hodgkin-Huxley model.

Here, C_m is the membrane capacitance, I_{input} represents the incoming current, and V is the membrane potential. When V exceeds a predefined threshold, an action potential is triggered, and the membrane potential is reset.

While the integrate-and-fire model sacrifices some of the biological detail of the HH model, it remains valuable for studying neuronal behavior in scenarios where computational efficiency is crucial. To see a typical response of a (leaky) integrate-and-fire model, let us look at 1.3.

This model is particularly well-suited for neuromorphic computing for several reasons. Firstly, its simplicity allows for efficient hardware implementation. The integrate-and-fire neurons can be represented using compact electronic components, making them suitable for large-scale integration in specialized neuromorphic hardware.

Additionally, the integrate-and-fire model aligns with the energy-efficient nature of neuromorphic computing. By focusing on the threshold-based firing mechanism, this model minimizes continuous power consumption and allows for efficient event-driven processing.

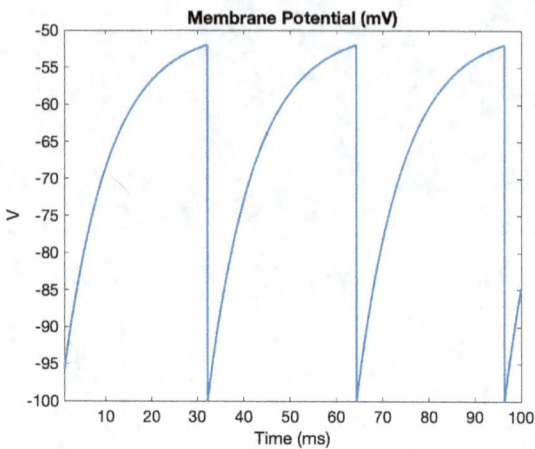

Figure 1.3: Spiking behavior in the leaky integrate-and-fire model. We can see the difference in the curves between the LIF and the HH model.

Furthermore, the integrate-and-fire model supports the concept of spike-based communication, which is a crucial feature of neuromorphic systems. Neurons communicate by emitting discrete spikes of information, similar to the way biological neurons transmit signals via action potentials.

The integrate-and-fire model has been a fundamental building block in neuromorphic computing due to its simplicity, efficiency, and compatibility with spike-based communication. Its ability to replicate essential aspects of neuronal behavior makes it a cornerstone in the development of energy-efficient and highly parallel neuromorphic systems. This is one of the important goals, currently, in the field of neuromorphic computing.

These types of neurons are also called spiking because of the response curves we have seen earlier. The question is then: can these devices be made in real life or are just theoretical? Let us now provide two examples of real neuromorphic devices.

Among the pioneers in this domain are two prominent neuromorphic devices: SpiNNaker and Loihi.

SpiNNaker: SpiNNaker, short for "Spiking Neural Network Architecture," is a neuromorphic computing platform developed by the University of Manchester in collaboration with the European Union's Human Brain Project. It is designed to simulate the behavior of large-scale neural networks with remarkable efficiency. SpiNNaker embodies a digital approach, utilizing a massively parallel array of ARM processors to mimic the spiking neural activity observed in biological brains, following a simulation of integrate and fire neurons.

With SpiNNaker, researchers can model and simulate networks comprising billions of spiking neurons in real-time. This platform is particularly adept at addressing challenges related to neuroscience research, cognitive modeling, and large-scale neural network simulations. Its digital nature allows for precise control and synchronization of neural activity, making it a valuable tool for scientific inquiry.

Loihi: Loihi, developed by Intel, is another leading neuromorphic computing device that employs a different architectural approach. Unlike SpiNNaker's digital emulation, Loihi is based on a mixed-signal design, incorporating both digital and analog elements. This unique configuration allows Loihi to execute spiking neural networks with exceptional energy efficiency, closely mirroring the biological brain's operation.

What sets Loihi apart is its capacity for on-chip learning. This means that the device can adapt and learn from new data without relying on external computations. This capability makes Loihi well-suited for applications involving adaptive learning, pattern recognition, and autonomous decision-making in real-world environments.

Key Differences between Loihi and SpiNNaker:

1. **Architecture:**

SpiNNaker: Digital, massively parallel array of ARM processors.
Loihi: Mixed-signal design, combining digital and analog elements.

2. **Learning Capability:**

SpiNNaker: Primarily designed for simulation and modeling, lacks on-chip learning.
Loihi: Equipped with on-chip learning capabilities, allowing for adaptive behavior.

Both SpiNNaker and Loihi represent significant strides in the field of neuromorphic computing, each offering distinct advantages and applications. While SpiNNaker excels in large-scale neural network simulations, Loihi's mixed-signal architecture and on-chip learning capabilities make it a powerful tool for adaptive, energy-efficient computing tasks.

Neuromorphic devices like Loihi and SpiNNaker offer a fundamentally different approach to computing compared to traditional CMOS (Complementary Metal-Oxide-Semiconductor) digital devices.

Thus, we have seen that the current generation of neuromorphic devices is explicitly designed to mimic the behavior of biological neurons and synapses. This allows them to process information in a manner more akin to the human brain, which is especially advantageous for tasks involving pattern recognition, learning, and adaptability. This is the reason for the name neuromorphic conventionally assigned to these types of architectures. Yet, these are highly engineered devices, using CMOS, rely on classical digital logic gates, and are not inherently designed to replicate neural processing at a fundamental level.

Neuromorphic devices like Loihi incorporate both digital and analog components. This mixed-signal approach enables them to perform computations in a manner more similar to the continuous, graded responses of biological neurons. CMOS devices operate exclusively in the digital domain, which relies on discrete levels (0 and 1) for computation.

Neuromorphic devices often exhibit superior energy efficiency compared to CMOS digital devices for certain types of computations, especially those related to neural processing. This is because they exploit the principles of spiking neural networks and mixed-signal processing, which can significantly reduce energy consumption for specific types of tasks. Neuromorphic devices inherently operate in a massively parallel fashion, which is well-suited for tasks involving large-scale neural network simulations. This allows them to process information in a way that is closer to how the brain functions. In contrast, while digital CMOS devices can be highly efficient for certain tasks, they may not exhibit the same degree of parallelism or real-time processing capabilities. Neuromorphic devices like Loihi have the potential to emulate certain forms of synaptic plasticity, allowing for learning and adaptation. This is a feature crucial for tasks that require continual refinement of models or adaptation to changing environments, which is not a native capability of traditional CMOS devices. Neuromorphic devices like Loihi and SpiNNaker offer a specialized approach to computing that is uniquely tailored for tasks involving neural processing, learning, and adaptability. Their mixed-signal architecture, on-chip learning, and energy-efficient operation set them certainly apart from traditional CMOS digital devices, making them particularly well-suited for a wide range of applications in artificial intelligence and neuroscience research.

Neural plasticity is the brain's remarkable ability to reorganize and adapt, is a fundamental aspect of learning and memory. Through a series of simple yet powerful examples, we will delve into the fascinating world of neural plasticity, starting with one of the most iconic experiments in psychology - Pavlov's dog. In the early 20th century, Russian physiologist Ivan Pavlov conducted experiments with dogs to explore the phenomenon of classical conditioning. Initially, he observed that dogs naturally salivate in response to food. This reflex is an innate, unconditioned response to an unconditioned stimulus.

Pavlov then introduced a neutral stimulus, such as a bell, before presenting the food. After repeated pairings of the bell and food, the dog began to associate the bell with the impending meal. Eventually, the dog would salivate at the sound of the bell alone, even in the absence of food. This learned response, now triggered by a conditioned stimulus, demonstrated a profound example of neural plasticity. At the neural level, this process involves changes in synaptic connections. Initially, the neural pathways associated with the bell were weak. However, through repeated pairings with the food, these connections strengthened, resulting in a new association between the bell and the salivation response. This synaptic plasticity is a hallmark of learning and memory formation. The concept of neural plasticity extends beyond Pavlov's dog and has significant implications in various aspects of human life. For example, rehabilitation after brain injuries often relies on the brain's ability to rewire itself. Through targeted exercises and therapies, individuals can regain lost functions by creating new neural pathways.

In neuromorphic computing, replicating classical experiments like Pavlov's dog offers a unique window into the world of artificial intelligence and neural processing. In a basic neuromorphic setup, we have two components: an input stimulus (represented as a digital signal) and a response generator (analogous to the salivation response in Pavlov's dog). The objective is to establish an association between the stimulus and response. The typical procedure to reproduce Pavlov's dog is simple.

One first starts by presenting the digital stimulus to the input component. This could be a visual or auditory signal. Then, the experimenter records the output of the response generator. Initially, this output is random and unconditioned.

The experiment is repeated several times. Repeatedly pair the stimulus with the response, similar to Pavlov's approach. Over time, the response generator should start producing the desired output in anticipation of the stimulus.

The next step is verification. The experimenter verifies the association by presenting the stimulus without any immediate reward. If the response generator generates the expected response, conditioning has been achieved.

The second set of experiments follows the first. The follow-up experiment explores the concept of extinction, where conditioned responses gradually fade when the stimulus is no longer followed by a reward.

Let us now see how neuromorphic devices as Loihi and SpiNNaker can reproduce Pavlov's dog.

To replicate Pavlov's dog experiment using neuromorphic devices, we can utilize specialized hardware designed to mimic the behavior of biological neurons and synapses. Two prominent neuromorphic devices that can be employed for this purpose are the SpiNNaker and Loihi neuromorphic processors, described earlier.

SpiNNaker can be programmed to represent the digital signals as stimuli and emulate the behavior of neurons in response to those signals. The platform enables the establishment of connections between neurons, simulating the associative learning process observed in Pavlov's experiment. By adjusting the synaptic weights in the SpiNNaker network, one can model the strengthening of connections over time.

Loihi can also be programmed to represent the neural responses to stimuli. It allows for the simulation of learning and conditioning processes observed in classical experiments. Similar to SpiNNaker, Loihi allows for the adjustment of synaptic weights, which is crucial for modeling the process of associative learning. For both SpiNNaker and Loihi, the process of setting up the experiment would involve programming the network architecture, defining the stimuli, and monitoring the responses over multiple trials. The neuromorphic devices would then adapt and modify their behavior based on the repeated presentation of the stimuli, emulating the process of classical conditioning seen in Pavlov's dog.

It is worth noting that while these neuromorphic devices can replicate the basic principles of Pavlov's experiment, they are still simplifications of complex biological systems. They provide a powerful platform for studying and simulating neural processes, but may not perfectly mimic all aspects of biological neural networks.

Now, it is important to note that Loihi and SpiNNaker are highly complex neuromorphic devices when it comes to their engineering. Pavlov's dog can however be reproduced with a much simpler device, the memristor.

In electronics, the memristor is a component that emerged over the last decade, promising to redefine the boundaries of computing and artificial intelligence. The memristor, a portmanteau of "memory" and "resistor", was conceived at a theoretical level by Leon Chua in the '70s. It took decades for this passive circuit element to materialize in the laboratory, although technically it had been known since the '60s in some materials such as oxides.

In order to introduce the memristor, let us first recall a few important points of electronics. Ohm's law is a fundamental principle in electronics of macroscopic conductors, that describes the relationships between voltage, current, resistance, and charge in a circuit. These laws provide essential insights into how electrical components, like resistors, behave. Voltage, often denoted as V, is the electric potential difference between two points in a circuit. It is measured in volts (V) and represents the force that drives the electric current. In simple terms, it is the "push" that propels electrons through a conductor. Current, represented as I, is the flow of electric charge through a conductor over time. It is measured in amperes (A). Think of it as the rate at which electrons pass a specific point in the circuit. Again, if you are not mathematically inclined, you can skip what follows or read at a high level without focusing on the details.

The resistance, denoted as R, is a property of a material that hinders the flow of electric current. It is measured in Ohms (Ω). A higher resistance means it is more difficult for electrons to pass through a material. Ohm's law states that the voltage across a resistor (V) is directly proportional to the current (I) flowing through it, given a constant resistance (R). This can be mathematically represented as:

$$V = I \cdot R$$

This means that the voltage drop across a resistor is equal to the current passing through it multiplied by the resistance.

In more advanced discussions, the concepts of flux (Φ) and charge (q) come into play. Flux refers to the flow of energy or particles through a surface, while charge represents the quantity of electrons in motion. This is for instance the case in "ideal" memristors. This is a distinction we will drop later, but it serves well in this introduction to the memristor. At a very basic level, we can think of a memristor as a resistor.

The memristor's resistance, R_m, is a function of the charge, $q(t)$, that has passed through it:

$$R_m = R_0 + \alpha \int_{-\infty}^{t} i(\tau)\, d\tau = R_0 + \alpha(q(t) - q(-\infty))$$

Here, R_0 is the initial resistance, α influences resistance change, and $i(t)$ is the current through the memristor. The memristor's behavior can also be expressed in terms of the flux, $\Phi(t)$, linked with the device: $\Phi(t) \propto \int_{-\infty}^{t} v(\tau)\, d\tau$ The rate of change of flux with respect to charge defines the memristance, denoted by M: $\frac{d\Phi}{dq} = \frac{1}{\alpha} = M$ This characterizes the relationship between charge and flux. A memristor exhibits a pinched hysteresis loop in the v-i plane due to its history-dependent behavior.

In realistic experiments, however, this is not as simple as that. In fact, when the voltage is removed, the memristor returns to its original non-conducting state, at a high resistance (or low conductance) value.

A simple model describes this quite effectively. We introduce a certain parameter $W(t)$ which characterizes the resistance state as a function of time, i.e. $R(t) \equiv R(W(t))$.

Then, we assume the existence of an equation for R and the evolution of W, e.g.

$$R(W) = R_{ON} * W + R_{OFF} * (1 - W) \tag{1.1}$$

$$\frac{dW}{dt} = \alpha I(t) - \beta W(t) \tag{1.2}$$

which we now explain. Above, W is constrained (somehow) between 0 and 1. In the literature, this is done in a variety of ways depending on which mechanisms underlie the material's resistance change.

If we apply a certain voltage, then (assuming $\alpha > 0$) the parameter W increases, meaning that the resistance is driven towards R_{ON}, which is assumed to be a low resistance state. Otherwise, if there is no voltage applied, the parameter W decays exponentially (controlled by β) to $W = 0$, and the resistance goes up to the R_{OFF} state. This is called also a switch sometimes. If β is not zero, then the memristor loses memory, and the parameter W is not exactly the charge anymore. This is a technical detail, but in the literature, there has been a fierce battle over this point. In fact, in Chua's original idea, this was supposed to be a function of the charge. The mathematically inclined readers can check that for $\beta = 0$, W must be proportional to the charge!

To see what happens with this simple rule, check for instance the behavior of this memristor when a positive or negative voltage is applied in Fig. 1.4

A memristor's non-volatile nature allows it to retain resistance levels even without power. This property enables it to store information, making it ideal for memory technology and neuromorphic computing. Neuromemristive systems (see for instance the Wikipedia page for Neuromorphic engineering) are made of a large number of memristors, and toy models for networks of these components have been proposed in the literature.

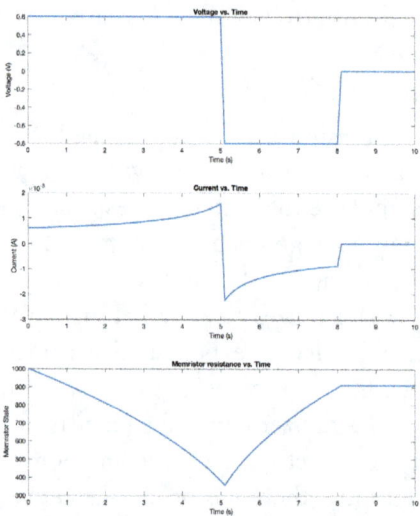

Figure 1.4: Evolution of the memristor's resistance as a function of the applied voltage.

The memristor's dynamic resistance behavior finds applications in memory storage, synaptic emulation in artificial neural networks, analog computing, and adaptive systems.

At its core, a memristor is an intriguing electronic element capable of retaining information in a non-volatile manner, akin to how synapses function in biological neural networks. Unlike conventional resistors, which merely impede the flow of current, memristors embody memory within their electrical resistance, exhibiting a transformative dual nature.

The memristor's versatility transcends traditional electronic components. Its unique ability to "remember" past electrical states position it at the forefront of applications ranging from non-volatile memory storage to neuro-inspired computing paradigms. This malleability, combined with its potential to perform logic operations, hints at a future where computational architectures seamlessly integrate memory and processing.

By mimicking the behavior of biological synapses, memristors can enable the creation of artificial neural networks with unprecedented efficiency and potential for learning and adaptation. This intrinsic capacity to modulate resistance based on prior electrical activity aligns seamlessly with the plasticity observed in biological synapses.

As we have seen, in biological organisms, synaptic plasticity is the cornerstone of learning and memory. It entails the modification of synaptic strength based on repeated neural signaling. The memristor's ability to emulate this synaptic plasticity on a silicon substrate opens doors to a new era of machine learning and artificial intelligence. The question is: can be used to emulate Pavlov's dog response?

Pavlov's classical conditioning experiment, which involves associating a neutral stimulus (like a bell) with a natural response (like salivation to food), primarily relies on the concept of neural plasticity. Neural plasticity involves changes in the synaptic connections between neurons, allowing them to adapt and form new associations. Memristors can emulate certain aspects of synaptic plasticity by adjusting their resistance based on electrical signals. This behavior is reminiscent of the changes in synaptic strength seen in biological neural networks during learning and memory processes. To replicate Pavlov's dog experiment, you would need a complete neural network model that includes sensory inputs, processing units (neurons), and the capacity to adjust synaptic weights (like memristors) based on the conditioning process. Creating such a model with memristors would be complex and likely require additional components. Pavlov's experiment involves not only synaptic changes but also the entire sensory-perception-learning-response loop of a biological organism. Memristor-based systems, while capable of simulating synaptic plasticity, do not have sensory organs, a brain, or the complexity of a living organism. Building a memristor-based system to replicate classical conditioning would require advanced hardware design, programming, and potentially a combination of other components (sensors, actuators) to create a full experimental setup. However, we can make some assumptions that will clarify some of the points behind neuromorphic systems.

2 Information theory and cybernetics

> The automatic machine, whatever we think of it, is one of the great figures of the Industrial Revolution. It is even more significant to us than the railway locomotive of the last century. We live in the machine age, and it is impossible to live without them.
>
> Norbert Wiener, from his book "The Human Use of Human Beings" (1950)

Claude Shannon's Theory of Information, developed in the mid-20th century, had a profound impact on our understanding of communication, leading the way for advancements in fields ranging from telecommunications to computer science.

Before Shannon's groundbreaking work, information was a nebulous concept, lacking a precise mathematical framework. Early discussions on information centered on philosophical and linguistic perspectives. For instance, philosophers like John Locke and George Berkeley explored the notion of information in terms of the conveyance of ideas or knowledge.

In the late 19th and early 20th centuries, the emergence of telecommunications brought about a new context for discussions on information. Engineers and inventors were grappling with how to transmit signals effectively and reliably over long distances. This led to the development of concepts like telegraphy and telephony, where the focus was on the efficiency of transmitting messages.

However, it was Claude Shannon who provided the formal foundation for a quantitative theory of information. In his landmark paper "A Mathematical Theory of Communication" published in 1948, Shannon introduced key concepts such as entropy, noise, and channel capacity.

Shannon's theory of information marked a paradigm shift. He defined information in a precise, quantifiable manner, based on the reduction of uncertainty in a message. The concept of entropy, borrowed from thermodynamics, was adapted to measure the average amount of information contained in a message. This was a profound breakthrough, as it allowed for the precise analysis and optimization of communication systems.

Shannon's theory of information had far-reaching implications. It provided the theoretical framework for designing efficient and reliable communication systems, influencing the development of technologies like digital coding, error correction, and data compression. It also laid the groundwork for the field of information theory, which has found applications in diverse fields such as cryptography, computer science, genetics, and neuroscience.

However, Shannon's theory started from the idea of communication. At an abstract level, communication can be understood as a fundamental process that underlies the exchange of information, meaning, or signals between entities. It is a ubiquitous phenomenon that occurs in various forms across the natural world and within artificial systems.

At its core, communication involves a sender, a message, a medium, and a receiver. The sender encodes information into a format that can be transmitted, which is then conveyed through a chosen medium. The receiver decodes the message to extract the intended meaning or information. This process can occur through various channels, including verbal language, written text, visual cues, gestures, signals, or even symbolic representations in computer systems.

Communication is not confined to language alone. It encompasses a broader spectrum of modalities, including non-verbal cues, visual displays, auditory signals, and even chemical or biological signaling in certain organisms (such as slime molds or ants, as we have seen). Furthermore, communication can occur at different levels of complexity, ranging from simple exchanges of signals in biological systems to highly sophisticated information transfer in human societies and technological networks.

One critical aspect of communication is the notion of shared understanding. For effective communication to occur, there must be a common framework of meaning between the sender and receiver. This shared understanding can be influenced by cultural, linguistic, and contextual factors, which may introduce nuances and complexities to the communication process.

Importantly, communication is not always explicit or intentional. Implicit forms of communication can occur through body language, facial expressions, or even the design of an environment. Additionally, communication can be affected by noise or interference, which may distort or disrupt the intended message.

Communication is like a relay race where information is passed from one person to another. Imagine you have something you want to share with a friend. This thought you want to convey is the message. Now, you need a way to express it.

This is where the encoder comes in. The encoder is like your personal translator. It takes the thoughts in your mind and converts them into a form that can be sent out. For instance, if you're talking face-to-face, your mouth and body language serve as the encoder. If you're sending a text message, it's your fingers on the keyboard.

Then comes the channel. Think of this as the pathway your message travels through. It could be the airwaves if you're talking, or the internet if you're sending a text. This is the route your message takes to reach your friend. The channel can be faulty, and that is where probability theory comes into play.

Now, your message is on its way, but it is not in its final form yet. It needs to be converted back into something your friend can understand. This is where the decoder steps in. It's like your friend's personal translator. It takes the signals or words they receive and turns them back into the original message. If you're talking, their ears and eyes do the decoding. If it is a text, it's their brain processing the words.

So, in this relay of information, you start as the encoder, turning thoughts into a message. The channel acts like a bridge, carrying your message. Finally, your friend becomes the decoder, transforming the signals back into the original idea. In information theory, entropy is a measure of the amount of uncertainty or randomness in a message. It is like a gauge of how surprising or unpredictable a message is. If a message has high entropy, it means there is a lot of unpredictability, and it is harder to guess what comes next. On the other hand, if a message has low entropy, it's more predictable and less surprising.

Think of it like this: if you're sending a message where every letter is equally likely, like in a random string of letters, it has high entropy. But if you're sending a message where certain letters or words are much more likely to appear, like in a coherent sentence, it has lower entropy.

If you use a code with high entropy, it means that even if someone intercepts it, it will be very hard for them to understand because it is so unpredictable. But if you use a code with low entropy, it might be easier for someone to crack because there are patterns or clues that give away the meaning.

To make this statement mathematical, imagine again you are trying to send a secret message to a friend, but you want to make sure it's as secure as possible. You decide to use a code, but you want to be efficient about it. This is where entropy comes into play.

In information theory, entropy (H) is a measure of the amount of uncertainty or randomness in a message. Mathematically, it can be expressed as

$$H(X) = -\sum p(x) \log_2 p(x) \tag{2.1}$$

Where:

$H(X)$ is the entropy of the random variable X.

$p(x)$ is the probability of outcome x.

If you are sending a message where every letter is equally likely, like in a random string of letters, it has high entropy. But if you are sending a message where certain letters or words are much more likely to appear, like in a coherent sentence, it has lower entropy.

So, in information theory, entropy helps us measure and understand the level of surprise or uncertainty in a message. It's a crucial concept in designing codes, encryption, and communication systems to ensure that messages are both secure and efficient.

Information theory has far-reaching implications and connections to various fields of science.

In thermodynamics, entropy is a fundamental concept that measures the amount of disorder or randomness in a physical system. It quantifies the dispersal of energy within a system. Interestingly, Claude Shannon, the pioneer of information theory, drew an analogy between information theory and thermodynamics. Shannon named entropy, in fact, after Boltzmann's entropy under the suggestion of von Neumann.

This analogy led to a deeper understanding of the relationship between information and physical systems. It revealed that information could be considered as a physical quantity, akin to energy or entropy in thermodynamics. This realization has profound implications, especially in the context of understanding the nature of information processing and its connection to the physical world.

In the realm of statistical mechanics, entropy emerges as a fundamental concept that characterizes the behavior of physical systems composed of a large number of particles. Let's make the case for entropy as a measure of the number of states:

Consider a system with a large number of particles, such as gas molecules in a container. Each molecule can exist in a multitude of possible states: it can have different positions, velocities, and energies. The number of possible configurations or arrangements of all these particles is astronomically vast.

Entropy, in statistical mechanics, is intimately linked to this vastness of possibilities. It quantifies the number of microstates that correspond to a given macrostate of the system. A macrostate refers to a specific set of observable properties, like the total energy, volume, and pressure of the system.

For a macrostate with fixed energy (which is often the case in many practical situations), the number of microstates associated with it, known as the multiplicity (Ω), increases exponentially with the system's size. Mathematically, it is represented as:

$$S = k \ln \Omega$$

where S is the entropy, k is the Boltzmann constant, and Ω is the multiplicity. For systems at equilibrium, Shannon's entropy and Boltzmann's entropy agree.

As the number of particles in the system increases, the multiplicity explodes, and so does the entropy. This means that there are an incredibly large number of microscopic configurations that correspond to a particular macroscopic state. In other words, entropy quantifies the sheer multitude of ways in which the microscopic constituents of a system can be arranged while still preserving the same macroscopic properties.

Beyond thermodynamics, information theory has ties to fields like communication engineering, computer science, neuroscience, genetics, and linguistics. It plays a crucial role in data compression, error correction, cryptography, and the study of complex systems. It also has a role in biology.

In biological processes, entropy is intricately connected to the flow of energy. The second law of thermodynamics states that in any energy transformation, the total entropy of a closed system tends to increase. In biological organisms, this implies that energy transactions involve an inherent rise in entropy. Organisms absorb energy from their environment, utilizing it for various physiological functions, and subsequently release it as waste heat, contributing to the overall increase in system entropy.

Consider the processes of metabolism and homeostasis. Living systems exhibit an impressive capacity to maintain a state of relative order and stability, a phenomenon known as homeostasis. This requires a continual infusion of energy to counteract the natural tendency of systems to become more disordered. Entropy, in this context, offers insight into the delicate balance between order and chaos within a living organism.

In the realm of genetics, entropy manifests in the storage and transmission of biological information. The genetic code, represented by DNA sequences, constitutes a highly structured form of information. Maintaining and faithfully replicating this information during cell division demands an investment of energy and intricate processes that push against the innate drive toward disorder.

Evolutionary processes also bear relevance to entropy. Evolution is marked by the generation of diversity within living organisms, akin to an increase in the variety of forms and functions. This surge in diversity aligns with the concept of increased entropy within the biological system.

Zooming out to ecological systems, we find that entropy influences energy flow and nutrient cycling. As energy moves from one trophic level to another, a portion is inevitably lost as waste heat, contributing to the overall rise in entropy. Similarly, nutrient cycling, involving the constant transformation and redistribution of matter, further contributes to the overall increase in entropy within an ecosystem.

Finally, at the core of it all, biological systems are ceaselessly adapting to changes in their environment. This adaptation involves a continuous exchange of information and energy with the surroundings. Here, entropy provides a lens through which we can appreciate the intricate dance between order and disorder that characterizes the biological world.

Let us now look instead at the case of resistive systems like those described in the introduction.

Imagine you are transmitting an electrical signal through a wire to convey information. In this scenario, the wire can be viewed as a resistive system. Resistive systems, like electrical circuits, introduce resistance to the flow of current, which can lead to losses in the form of heat.

Now, consider a situation where you are sending a digital message using a binary code. In this code, information is represented by a sequence of binary digits (0s and 1s). Each binary digit corresponds to a different voltage level: for instance, 0 might represent a low voltage, and 1 might represent a high voltage.

When you transmit this digital signal through a wire, it encounters resistance. As the current flows through the wire, some of its energy is converted into heat due to the resistance. This leads to a phenomenon known as "thermal noise" or "Johnson-Nyquist noise."

Here is where entropy comes into play. Entropy, in this context, refers to the randomness or unpredictability introduced by this thermal noise. It adds an element of uncertainty to the received signal. This means that even if you transmit a perfectly clear message, by the time it reaches the other end, it may have some level of noise or distortion due to the resistance in the wire.

In communication systems, managing entropy is crucial for ensuring reliable transmission. Techniques like error correction codes are employed to detect and correct any errors introduced by noise. These codes add redundancy to the message, allowing the receiver to identify and fix errors caused by entropy.

So, in this example, entropy is significant in communication with resistive systems because it represents the degree of randomness or unpredictability introduced by the inherent resistance in the transmission medium. Managing and mitigating this entropy is essential for maintaining the integrity and accuracy of the transmitted information.

This notion of entropy provides a powerful framework for understanding the behavior of systems on a macroscopic scale based on the statistical properties of their microscopic constituents. It allows us to predict and explain phenomena like phase transitions, heat transfer, and the behavior of gases.

Our goal is to explain the connection between information and biological systems. The field of cybernetics had an important role in this respect. In fact, information theory and cybernetics share a profound connection, as they both revolve around the fundamental concepts of communication, control, and the processing of information within systems.

As we have seen, information theory, is concerned with quantifying information, defining measures of uncertainty (entropy), and understanding the efficient encoding, transmission, and decoding of messages. Cybernetics, on the other hand, delves into the broader study of systems, encompassing biological, mechanical, social, and cognitive entities, examining how they process information to maintain stability and achieve goals.

This intersection is particularly evident in the concept of feedback loops. In both information theory and cybernetics, feedback plays a crucial role. In information theory, feedback is essential for error correction and efficient communication. In cybernetics, feedback is the mechanism through which a system monitors its own output and adjusts its behavior to maintain stability and adapt to changing conditions.

The term "cybernetics" was coined by mathematician Norbert Wiener in his influential book "Cybernetics: or Control and Communication in the Animal and the Machine," published in 1948. Wiener drew upon ideas from various fields, including engineering, biology, and feedback control systems. Cybernetics aimed to bridge gaps between traditionally separate disciplines such as mathematics, biology, psychology, engineering, and social sciences. It emphasized the study of systems as a whole, rather than analyzing individual components in isolation. Alongside Norbert Wiener, other key figures in the development of cybernetics included mathematician John von Neumann, biologist Ludwig von Bertalanffy (known for his General Systems Theory), engineer Claude Shannon (known for information theory), and anthropologist Gregory Bateson.

One of the central concepts in cybernetics is feedback, which involves the process of a system receiving information about its own behavior and using it to adjust its operations. This concept played a crucial role in the development of control systems theory. Norbert Wiener was the first to take seriously the idea that feedback in dynamical systems is important.

A later development known as "second-order cybernetics" emerged in the 1970s. It focused on reflexivity and the observer's role in the system, acknowledging that the observer is an integral part of any system being studied.

We can simplify cybernetics as the science where "automata" attempt to achieve goals via feedback and information.

Norbert Wiener (in his book "The Human Use of Human Beings", 1950) envisioned intelligent behavior arising from the intricate interplay of feedback loops. He astutely observed these feedback processes, involving sensors, signals, and actuators, permeating all aspects of existence, from living systems to interactions between humans and machines. In the 1940s, Wiener's cybernetic theory of feedback, communication, and control proved instrumental in resolving various engineering challenges, finding applications in realms as diverse as assembly lines and rocketry.

Wiener was remarkably prescient, grappling with the profound societal and individual repercussions of technology. His perspective took a sad turn, harboring concerns over the potential use of machines for human control and job displacement. In his book, Wiener sounded a cautionary note against this bleak trajectory for humanity. He advocated for the advancement of technology that amplifies human capacities rather than exerting dominion over them. Specifically, Wiener expounded on how machines could emulate the "communicative mechanisms" of the human nervous system, potentially revolutionizing prosthetics and the restoration of human functions. These insights, though visionary, anticipated a gradual process, with the development of machines capable of meaningfully interfacing with neural signals taking several decades to materialize.

Norbert Wiener's pioneering work in cybernetics laid the foundation for several concepts and algorithms that served as precursors to artificial intelligence (AI). One of the key concepts was feedback loops, which formed the basis for control systems and adaptive processes. This notion of self-regulation and self-correction is fundamental to both cybernetics and AI.

Another important precursor was Wiener's development of signal processing techniques, which are essential in both cybernetics and AI for tasks like pattern recognition and data analysis. Cybernetics explores how systems, information, and control work together, it is not just about hardware and software, but also how natural and artificial systems interact. When thinking about future computer-brain cooperation, cybernetics broadens our perspective. It prompts us to consider not just the tech, but also the ethical and societal impacts. Things like privacy and autonomy are important, as well as the concept of consciousness. This is why it is important to mention at this stage of this book cybernetics. Cybernetics also encourages us to think about how these systems will learn and adapt over time, which is important for neuromorphic computers. It goes beyond just the nuts and bolts of technology to consider how human thinking and artificial intelligence can complement each other.

Thus, by including cybernetics in the conversation, we get a more well-rounded understanding of what the future might look like when computers and brains team up. It's not just about the tech, but also the big questions about how we'll navigate this new era of human-computer synergy.

Presently, many view cybernetics as a precursor to fields like Artificial Intelligence, Systems Theory and Complex Systems. While there is some validity to this perspective, it is important to acknowledge that these disciplines also have their own distinct and complex origins. Cybernetics, fundamentally, embodied more of a technological philosophy than a strict discipline. It provided a lens through which to perceive and approach communication, systems, control, and human-machine interaction, rather than a rigid set of technical principles and methodologies that engineers could directly apply to analyze or design systems or machines. In this regard, it often contended with alternative approaches such as Systems Dynamics, Systems Engineering, and Systems Analysis, each of which adopted a more focused scope, resulting in more tangible practices and outcomes. The goal of this section is to provide a cybernetic point of view on neuromorphic computing, explicitly showing how cybernetics has impacted various fields. Again, if you are not mathematically inclined, feel free to skip the details and read at a high level.

Control theory.

Control theory has found applications in a diverse array of fields. In aerospace engineering, it's used to stabilize aircraft and spacecraft, ensuring they maintain desired trajectories. In chemical engineering, it's employed to optimize processes and ensure consistent product quality. In robotics, it enables precise motion control and autonomous navigation. Let us now see this mathematically,

A dynamic system is typically described by a set of differential or difference equations. For continuous-time systems, the dynamics can be represented by a set of differential equations of the form:

$$\dot{x}(t) = f(x(t), u(t), t)$$

where $x(t)$ represents the state vector, $u(t)$ is the control input, and f is a vector-valued function describing the system dynamics. For discrete-time systems, the dynamics are described by the difference equation:

$$x[k+1] = g(x[k], u[k], k)$$

The primary objectives of control theory include stability, performance, and robustness.

Stability analysis involves studying the behavior of the system over time. A system is considered stable if, for any bounded initial condition and bounded input, the state trajectories remain bounded over time.

Performance measures quantify how well the system performs in achieving its desired objectives. These measures can include settling time, overshoot, rise time, and steady-state error.

Robustness addresses the system's ability to maintain desired performance even in the presence of uncertainties or variations in system parameters.

The design of a controller involves selecting an appropriate control law that determines the control input $u(t)$ or $u[k]$ based on the system's state and possibly other information (such as reference signals or disturbances).

Controllers can be designed using various techniques, including classical control methods like PID (Proportional-Integral-Derivative) control, as well as more advanced methods like state-space control, optimal control, and adaptive control.

Feedback control is a central concept in control theory. It involves continuously measuring the output of the system, comparing it to a desired reference signal, and using the error signal to adjust the control input. Mathematically, this can be expressed as:

$$u(t) = K \cdot (r(t) - y(t))$$

where K is the controller gain, $r(t)$ is the reference signal, and $y(t)$ is the system output.

Example 1: Temperature Control

Consider a heating system in a room. The temperature T is the state variable. The control input u is the power of the heater, which can be adjusted. The system output y is the measured temperature in the room. The reference signal r might be a desired temperature setpoint.

The dynamics can be represented by a simple first-order differential equation:

$$\dot{T}(t) = -a(T(t) - T_{\text{ext}}(t)) + b \cdot u(t)$$

where a and b are system parameters, and $T_{\text{ext}}(t)$ is the external temperature.

Example 2: Cruise Control

In an automotive cruise control system, the state variable x is the speed of the car. The control input u is the throttle setting. The system output y is the actual speed of the car. The reference signal r could be the desired speed set by the driver.

The dynamics can be described by a first-order differential equation:

$$\dot{v}(t) = -a(v(t) - r(t)) + b \cdot u(t)$$

where a and b are system parameters.

Example 3: Water Tank Level Control

Consider a water tank with an inlet valve and an outlet valve. Water enters the tank from the top at a rate proportional to a control parameter, for instance, the voltage V applied to a controlled valve or a pump. The water leaves through an opening in the tank base at a rate that is proportional to the square root of the water height, h, in the tank. The water level h is the state variable. The control input u is the opening of the inlet valve and it could represent the voltage. The system output y is the water level in the tank. The reference signal r could be a desired water level setpoint.

The dynamics can be represented by a differential equation:

$$\dot{h}(t) = c \cdot u(t) - d \cdot \sqrt{h(t)}$$

where c and d are system parameters.

Control theory, deeply rooted in cybernetics, provides a mathematical framework for understanding and manipulating dynamic systems. By harnessing feedback mechanisms and employing sophisticated control strategies, control theory continues to shape the advancement of technology and automation in various industries.

Perception. Warren McCulloch and Walter Pitts were pioneers in the field of neural network theory. In 1943, they published a seminal paper titled "A Logical Calculus of the Ideas Immanent in Nervous Activity," which laid the groundwork for understanding how simple computational units, inspired by the functioning of neurons, could be used to model perception and decision-making.

McCulloch and Pitts proposed a formal model of a simplified neuron, which they called a "threshold logic unit." This unit took binary inputs (either on or off) and produced a binary output based on a threshold.

The model was characterized by three key elements:

Inputs and Weights: The unit received inputs from other units or external sources, each of which was assigned a weight. These weights determined the importance of each input in the computation.

Aggregation: The weighted inputs were summed, and if the sum exceeded a certain threshold, the unit would produce an output signal (often represented as a binary "1"); otherwise, it would remain inactive (represented as a binary "0").

Activation Function: This function represented the threshold operation, determining whether the unit would fire based on the weighted inputs.

This threshold logic unit provided a foundational framework for understanding how simple computational elements could be organized into networks to perform complex tasks. By connecting these units in various configurations, McCulloch and Pitts showed that it was possible to create logical circuits capable of performing computations.

Their work was significant in demonstrating that networks of simple, binary units could, in theory, perform any computation that a digital computer could. This laid the groundwork for the development of artificial neural networks, which form the basis of many modern machine learning algorithms, especially in areas like pattern recognition, image processing, and natural language understanding.

We now provide a mathematical description of this model, which describes the functioning of a neuron. We consider a neuron represented by the McCulloch-Pitts model. It has n binary input signals $x_1, x_2, \ldots, x_n$ and corresponding weights $w_1, w_2, \ldots, w_n$. The neuron aggregates these inputs, computes a weighted sum, and applies an activation function to determine its output.

The weighted sum, denoted as z, is calculated as:

$$z = \sum_{i=1}^{n} w_i x_i$$

The neuron then applies a step function to z to produce the output y:

$$y = \begin{cases} 1 & \text{if } z \geq \theta \\ 0 & \text{otherwise} \end{cases}$$

where θ is a threshold parameter.

This model can be succinctly represented using the Heaviside step function:

$$y = H(z - \theta)$$

where $H(\cdot)$ is a certain threshold function.

The McCulloch-Pitts model provides a binary output, mimicking the behavior of biological neurons. It's important to note that this model operates in a feedforward manner, with information flowing from inputs to outputs.

The significance of the McCulloch-Pitts model lies in its demonstration that networks of these simple, binary units can perform complex computations. It paved the way for the development of more sophisticated neural network architectures, which have become instrumental in various machine learning applications.

In the McCulloch-Pitts model, nonlinearity plays a crucial role in determining the firing behavior of a neuron. The nonlinearity arises from the activation function applied to the weighted sum of inputs.

The activation function in the McCulloch-Pitts model is a step function (specifically, the Heaviside step function), which introduces a discontinuity at the threshold θ. This means that the output of the neuron abruptly switches from 0 to 1 (or vice versa) once the weighted sum of inputs surpasses the threshold.

In the extended McCulloch-Pitts model with a Rectified Linear Unit (ReLU) activation function, nonlinearity still plays a vital role in determining the firing behavior of a neuron.

The ReLU function is defined as:

$$H(z) = \text{ReLU}(z) = \max(0, z)$$

In this model, z is the weighted sum of inputs. The ReLU function introduces a nonlinearity that is more continuous compared to the step function used in the original McCulloch-Pitts model.

The ReLU function allows the neuron to be more sensitive to variations in input, especially when the weighted sum of inputs is greater than zero. This means that the neuron can respond to a wider range of input patterns, as it doesn't have a strict binary output. Instead, it allows for a graded response that increases with the strength of the input.

The introduction of the ReLU function in the McCulloch-Pitts model enhances its capacity for gradient-based learning algorithms, which are widely used in modern deep learning networks. This nonlinearity is crucial for enabling the neuron to capture complex relationships between inputs and outputs.

In another model, the hyperbolic tangent (tanh) activation function, nonlinearity continues to play a crucial role in determining the firing behavior of a neuron.

The tanh function is defined as:

$$H(z) = \tanh(z) = \frac{e^z - e^{-z}}{e^z + e^{-z}}$$

In this model, z represents the weighted sum of inputs. The tanh function introduces a smooth, S-shaped nonlinearity that allows for a more continuous transition between the neuron's output states.

Unlike the step function, which provides a binary output, the tanh function allows for a graded response. This means that the neuron's output can take on a range of values between -1 and 1, depending on the strength of the input.

The introduction of the tanh function in the McCulloch-Pitts model enhances its capacity to capture more complex relationships between inputs and outputs. This smooth nonlinearity is particularly advantageous in tasks where a continuous, smooth transition between states is desirable.

Nonlinearity introduces a critical property to the model. Without it, the McCulloch-Pitts model would essentially reduce to a linear model, and the neuron's behavior would be entirely determined by the weighted sum of inputs. In that case, multiple linear combinations of inputs would simply result in another linear combination, limiting the complexity of computations that can be performed.

Cybernetics and biology. At its essence, cybernetics in biology represents a paradigm shift in how we understand and model living systems. It introduces a framework that views organisms as dynamic, self-regulating entities engaged in a continuous dance with their environment.

Rather than viewing biological processes in isolation, cybernetics emphasizes the intricate web of feedback loops and information flows that govern the behavior of living organisms. It delves into the mechanisms by which organisms gather and process information, make decisions, and adapt to changing conditions. This perspective has revolutionized our understanding of phenomena ranging from cellular signaling pathways to the behavior of complex ecosystems.

In this context, cybernetics provides a powerful lens through which we can explore concepts like homeostasis, adaptation, and evolution. It illuminates the ways in which living systems maintain stability in the face of dynamic environments, and how they evolve and optimize their responses over time.

To understand this better, let us focus on homeostasis, which is a fundamental principle in biology that refers to the ability of a living organism to maintain stable internal conditions in the face of changing external environments. It's akin to a biological thermostat that ensures the body's internal environment remains within a narrow range, conducive to optimal functioning.

This concept was first introduced by the French physiologist Claude Bernard in the 19th century and later popularized by Walter Cannon.

The term itself is derived from the Greek words "homeo," meaning "similar," and "stasis," meaning "standing still." In essence, homeostasis describes the dynamic equilibrium that living organisms constantly strive to achieve.

The key components involved in homeostasis include sensors (receptors), control systems (often the nervous or endocrine systems), and effectors (muscles or glands that respond to the signals from the control systems).

For example, consider the regulation of body temperature in a human. When the external environment becomes colder, the body's temperature sensors detect this change and send signals to the brain's control center. In response, the brain triggers mechanisms such as shivering to generate heat and vasoconstriction to reduce heat loss. Conversely, if the environment becomes too warm, the body activates mechanisms like sweating and vasodilation to cool down. Incidentally, vasodilation is also the important mechanism behind the evolution of slime molds we discussed earlier.

Similarly, homeostasis governs a wide array of physiological parameters, including blood glucose levels, pH balance, blood pressure, and concentrations of ions like sodium and potassium. It's crucial for the proper functioning of cells, tissues, and organ systems. Without homeostatic mechanisms, the internal environment would be subject to erratic fluctuations, making it difficult for living organisms to carry out essential biological processes.

Importantly, homeostasis is not a static state, but a dynamic process. It involves continual adjustments and feedback loops that maintain internal stability in the face of ever-changing external conditions. This adaptability is a hallmark of living systems and a testament to their remarkable resilience.

The next biological feature we wish to discuss is the concept of adaptation, as it plays a role in homeokinesis. Adaptation is a central concept in biology that refers to the process by which organisms evolve and adjust to their environment over time. It is a fundamental mechanism that enables living beings to survive, reproduce, and thrive in diverse ecological niches. Adaptation occurs at various levels, from the molecular and physiological to the behavioral and anatomical. This was important for our introduction to Pavlov's dog. Connecting this to biology, Pavlov's experiments illustrate how organisms can adapt their behaviors and responses based on their experiences and the information they receive from their environment. This type of adaptive learning is crucial for survival, as it enables animals to anticipate and prepare for significant events, such as finding food or evading predators.

Expanding on this concept, the McCulloch-Pitts model, although initially focused on artificial neurons, also holds relevance in understanding adaptation in biological systems. The model's representation of neurons as binary units, where an output is triggered if a certain threshold is reached, mirrors the simplified way in which neurons transmit signals in real biological systems. While the McCulloch-Pitts model is a simplified representation, it captures the essence of how neurons process information and respond to stimuli.

In biological neurons, adaptation is a complex phenomenon that involves intricate processes such as synaptic plasticity and modulation of neurotransmitter release. These mechanisms allow neurons to adjust their responsiveness to incoming signals, a crucial aspect of learning, memory, and overall neural function.

Homeokinesis is a concept that shares similarities with homeostasis but differs in some key aspects, incorporating adaptation. While both terms refer to processes that help maintain stability within a living organism, they operate in slightly different ways.

Homeostasis, as previously explained, involves actively regulating specific physiological parameters (such as temperature, pH, or blood glucose levels) within a narrow range to ensure optimal functioning. This regulation is achieved through feedback mechanisms that counteract any deviations from the set point.

Homeokinesis, on the other hand, is a broader concept that encompasses the overall dynamic stability of an organism in response to a constantly changing environment. Rather than focusing on specific physiological parameters, homeokinesis emphasizes the ability of an organism to adapt and respond to various environmental conditions in a way that promotes its overall well-being and survival.

In homeokinesis, the emphasis is on flexibility and adaptability rather than strict regulation of specific variables. It acknowledges that the environment is not static, and organisms must be capable of adjusting their behavior, physiology, and other aspects of their biology to thrive in different conditions.

For example, an organism exhibiting homeokinesis might alter its activity levels, foraging behaviors, or reproductive strategies in response to seasonal changes, food availability, or predation risk. This allows the organism to maintain a general state of stability or equilibrium in the face of varying external circumstances.

Another important aspect of biology is evolution. Evolution stands as one of the most profound and foundational concepts in the field of biology. It represents the overarching framework that explains the diversity of life on Earth and how living organisms have changed over vast stretches of time. At its core, evolution is the process by which populations of organisms undergo gradual, heritable changes in their characteristics, leading to the emergence of new species and the modification of existing ones.

The idea of evolution was formalized by Charles Darwin in the mid-19th century, though similar notions had been suggested by earlier naturalists. Darwin's groundbreaking work, "On the Origin of Species," introduced the concept of natural selection as the primary mechanism driving evolution. Natural selection is the process by which individuals with advantageous traits for their environment tend to survive and reproduce, passing those traits on to their offspring. Over time, this leads to the accumulation of beneficial characteristics within a population.

Evolution operates on multiple levels, from the microevolutionary scale involving changes within populations, to the macroevolutionary scale involving the divergence and speciation of different lineages. It encompasses a wide range of mechanisms, including genetic drift, mutation, migration, and selective pressures from the environment.

The evidence for evolution is extensive and comes from diverse sources, including the fossil record, comparative anatomy, molecular biology, and biogeography. However, evolution today is a theory that has evolved itself, from Darwin's original. Fossil records provide a chronological record of past life forms, showcasing the transition from ancient species to their modern descendants. Comparative anatomy reveals shared structures and genetic information among different species, reflecting their common ancestry. Molecular biology, through the study of DNA and genetic sequences, offers insights into the relatedness of species at the genetic level.

Understanding evolution not only sheds light on the history and interconnectedness of all life forms but also has practical implications in fields like medicine, agriculture, and conservation biology. It allows us to make informed decisions regarding species preservation, disease management, and the development of new technologies. In particular, evolution provides a new perspective when it comes to generations of new computing architectures and chips.

In particular, evolution and cybernetics, seemingly distinct disciplines, share intriguing parallels that reveal deep insights into the functioning of both natural and technological systems.

At the heart of both fields lies the concept of feedback. In evolution, the process of natural selection functions as a feedback mechanism. Organisms with advantageous traits thrive and pass those traits to successive generations, shaping the genetic makeup of populations. Similarly, in cybernetics, feedback loops are fundamental to control systems. They allow for continuous adjustment and regulation in response to changing conditions, ensuring stability and efficiency in engineered systems.

Adaptation, a cornerstone of both evolution and cybernetics, demonstrates how systems, whether biological or artificial, respond to their environments. In evolution, organisms adapt over generations through genetic variations that confer advantages in specific contexts. In cybernetics, systems learn and adjust based on input and feedback, demonstrating a capacity for adaptive behavior.

The emergence of complexity from simplicity is another shared characteristic. Evolution illustrates how intricate biological structures and behaviors arise from incremental genetic changes. Cybernetics similarly demonstrates how complex behaviors can emerge from relatively straightforward feedback mechanisms, exemplified by the sophisticated capabilities of artificial neural networks and control systems.

Both fields heavily rely on modeling and simulation. Evolutionary models allow researchers to simulate how populations evolve over time, providing valuable insights into the dynamics of biodiversity. In cybernetics, modeling is essential for predicting and understanding the behavior of systems in response to various inputs and feedback.

Furthermore, the reciprocal influence between biology and technology is evident. Evolutionary principles have inspired the development of powerful optimization algorithms used in various engineering applications. Meanwhile, concepts from cybernetics, such as feedback control systems, have been instrumental in designing autonomous robots and intelligent systems capable of adapting to dynamic environments.

Imagine a scenario where a thermostat in your home is maintaining a comfortable temperature. It does this by constantly receiving information about the current temperature and adjusting the heating or cooling accordingly. This continuous exchange of information, known as a feedback loop, is fundamental to how many systems operate.

Now, let us discuss how information theory comes into play. In the context of feedback loops, information theory provides us with a structured way to think about and quantify the information exchange within the loop.

One of the key aspects is the ability to measure the amount of information being transmitted. This is crucial for ensuring that the information received is accurate and reliable. Information theory gives us tools to assess the effectiveness of this information flow, allowing us to optimize the communication channels.

Additionally, feedback loops often deal with repetitive or redundant information. Information theory offers techniques to manage this redundancy, ensuring that only essential information is transmitted. Think of it as a way to streamline the communication process, reducing unnecessary data while retaining critical content.

Noise and disturbances are common challenges in feedback systems. Information theory helps us design mechanisms to detect and correct errors introduced by noise. By doing so, we can minimize the impact of disturbances on the stability and performance of the system.

In adaptive systems, feedback loops play a crucial role in learning and adaptation. Information theory guides us in finding the right balance between exploring new information and exploiting existing knowledge. This ensures that the system can adapt and respond effectively to changes in its environment.

Furthermore, information theory helps us understand and manage information delays. In a feedback loop, there may be a delay in the transmission of information. By characterizing these delays, we can evaluate their impact on the system's behavior and take measures to mitigate any adverse effects.

Predictive modeling is another area where information theory shines. By leveraging techniques like entropy, we can build models that forecast future states of the system based on available feedback. This predictive capability is essential for making informed decisions and proactively adjusting system parameters.

However, entropy plays an important role in cybernetics. Although there have been a number of important developments in cybernetics, we mention one of these, which is called the Law of Requisite Variety formulated by W. Ross Ashby. It states that for a system to effectively control or regulate another system, it must possess at least as much variety as the system it is trying to control. The word variety, here, can be paralleled to a notion of "internal complexity". The Law of Requisite Variety essentially states that for a system to effectively regulate and control its environment, it must possess a degree of internal complexity or variety that matches or exceeds the complexity of the external environment it is interacting with. In simpler terms, it emphasizes that a system needs a sufficient range of responses or capabilities to effectively manage and adapt to the diversity and unpredictability of its surroundings. Entropy, in information theory, is a measure of the amount of disorder or uncertainty in a system. It quantifies the average amount of information needed to describe the state of a random variable or system. In simpler terms, higher entropy indicates greater randomness or unpredictability.

Variety, on the other hand, relates to the number of distinct states or responses available to a system. It signifies the diversity of options a system has for interacting with its environment. The connection between variety and entropy lies in their relationship to information. In a system, a higher variety means there are more possible states or responses available. This can lead to increased uncertainty or unpredictability in the system's behavior. In terms of information theory, higher variety can lead to higher entropy because describing the state of a system with greater variety requires more information. Conversely, if a system has low variety, it means there are fewer possible states or responses, leading to lower uncertainty or entropy. Describing the state of such a system requires less information.

Mathematically, if $V(R)$ represents the variety of the regulator and $V(S)$ represents the variety of the system, the law can be expressed as:

$$V(R) \geq V(S)$$

In practical terms, this means that the regulatory mechanism must be at least as complex as the system it regulates. If the regulator's variety is less than the system's variety, it won't be able to adequately respond to the system's changes and disturbances.

The Law of Requisite Variety underscores the importance of adaptability and complexity in control systems, emphasizing the need to match the regulatory capacity to the complexity of the system being controlled.

Consider a cruise control system in a car. The car's speed is the system being controlled. The cruise control mechanism needs to have a variety of possible adjustments to effectively regulate the car's speed. If the cruise control only had a single fixed speed setting, it would not have the requisite variety to control the car's speed under different conditions.

In the human body, the regulation of blood sugar levels is a complex process. The pancreas acts as a regulator, producing insulin to lower blood sugar levels and glucagon to raise them. The variety of possible insulin and glucagon levels needs to be sufficient to effectively respond to various changes in blood sugar levels, such as after a meal or during physical activity. If the regulator (pancreas) couldn't produce a range of insulin and glucagon levels, it would not have the requisite variety to maintain stable blood sugar levels.

Let us consider a feedback control system in an electrical circuit. The goal is to maintain a stable output voltage despite variations in load resistance.

The regulator (feedback control circuit) needs to be able to adjust the input voltage based on the output voltage to counteract changes in load resistance. If the regulator only had a limited range of adjustments or lacked the necessary complexity to respond to different load conditions, it would not have the requisite variety to effectively regulate the output voltage.

In this context, the Law of Requisite Variety emphasizes the importance of having a regulator with enough complexity to handle the range of conditions and disturbances that the electrical circuit may encounter.

Cybernetics, in general, was extremely important for the development of control theory. Control theory is a branch of mathematics and engineering concerned with designing systems that regulate the behavior of dynamic processes. It plays a pivotal role in a wide range of applications, from automated industrial processes to advanced aerospace systems. At its core, control theory seeks to understand how to manipulate inputs to a system in order to achieve desired outputs, despite uncertainties and disturbances.

The foundations of control theory find deep roots in the interdisciplinary field of cybernetics. Cybernetics, a term coined by Norbert Wiener, is the study of systems, control, and communication in animals, machines, and organizations. It emerged in the mid-20th century and brought together concepts from mathematics, engineering, biology, and philosophy. Central to cybernetics is the idea of feedback, where a system's output is used to adjust its behavior, creating a closed-loop control system.

The mathematical foundation of control theory lies in differential equations, linear algebra, and systems theory. These mathematical tools allow us to model the behavior of dynamic systems and analyze their responses to different inputs. Control systems are often represented using transfer functions or state-space models, which provide a concise mathematical description of the system's dynamics.

A key concept in control theory is stability. A system is considered stable if it returns to a desired state after being subjected to disturbances or changes in input. Stability analysis involves studying the poles of the system's transfer function or the eigenvalues of its state-space representation.

Feedback is a fundamental principle in both control theory and cybernetics. It involves continuously measuring the output of a system and using that information to adjust the system's inputs. This closed-loop control mechanism allows for automatic adjustments in response to changing conditions, making it a powerful tool in maintaining stability and achieving desired performance.

The Law of Requisite Variety, a cornerstone of cybernetics, asserts that in order for a system to effectively control another system, it must possess at least as much variety as the system it is regulating. This principle places emphasis on the need for complexity and adaptability in control systems.

Cybernetics is however not only about the mathematical aspects of the control of the system, but also about design principles.

Let us make a few more examples. In biological systems, for instance, we are fundamentally concerned with comprehending the intricate interplay of components. Biological systems are typical examples of complex adaptive systems. These could be cells, interacting with each other in a variety of ways, or ecological systems. It is often imperative first to meticulously account and deeply understand the pathways through which information propagates within the system.

Simultaneously, it is essential to adopt a dual perspective, scrutinizing the system at both microscopic and macroscopic levels. By examining the intricate workings of constituent elements, we gain insight into the holistic functioning of the system.

Moreover, one must not disregard the innate adaptability of biological systems. Akin to an organism's capacity to respond and acclimate to shifting circumstances, it is imperative to incorporate mechanisms that accommodate alterations over time.

In cybernetics, one incorporates these ideas and thinks about user-focused control. For medical practitioners, precision lies in identifying aberrations within the system, particularly in the context of pathological conditions. Furthermore, it is pertinent to evaluate the influence of diverse treatment modalities on the biological system. This involves simulating the impact of interventions, akin to employing a repertoire of strategies to ascertain potential outcomes. This necessitates a focused examination of areas exhibiting deviations from the norm.

On the other hand, ecologists adopt the role of investigators, deciphering the intricate web of interactions within ecosystems. This holistic approach involves scrutinizing the synergies between diverse species.

One needs to account for environmental perturbations, such as alterations in (for instance) temperature or habitat loss. This entails understanding how external factors introduce unpredictable variables into the system's dynamics.

3 The real brain and some brain models

Trying to understand perception by studying only neurons is like trying to understand bird flight by studying only feathers; it just cannot be done. To understand bird flight, you need to understand aerodynamics, only then can one make sense of the structure of the feathers and the shape of the wings. Similarly, you cannot reach an understanding of why neurons in the visual system behave the way they do, just by studying their anatomy and physiology.

David Marr (1982)

The most complex and enigmatic organ in the human body is the brain. Nestled within the protective confines of the skull, the brain is a marvel of nature, comprising an intricate network of cells and connections that orchestrate our thoughts, emotions, and actions.

At its core, the brain is composed of billions of specialized cells called neurons, each with the remarkable ability to transmit electrical signals. The model we have seen earlier should be considered now as a simplistic description of how the brain transmits and encodes information. There are various types of neurons, but these form an extensive web of communication, creating an astonishing 100 trillion synaptic connections. If stretched out, these connections could wrap around the Earth not once, but over four thousand times. That's the kind of complexity we are dealing with.

The brain's architecture is divided into distinct regions, each with its own specialized functions. The cerebral cortex, often referred to as the brain's "thinking cap," is responsible for higher cognitive processes like reasoning, perception, and voluntary movements. It is divided into lobes, including the frontal lobe (the seat of decision-making) and the temporal lobe (home to our auditory processing center).

Deep within the brain lies the fascinating limbic system, often called the "emotional brain." This region governs our feelings, motivations, and memories, giving color and depth to our experiences. The amygdala, a key player in emotional responses, is named after the Greek word for almond.

The cerebellum, often likened to a "mini-brain," handles coordination, balance, and fine motor skills. It houses an impressive 69 billion neurons, more than all the neurons in the rest of the brain combined. This little powerhouse ensures that you can dance, ride a bike, or even perform intricate tasks like playing a musical instrument.

And let us not forget the brainstem, the unsung hero that regulates life-sustaining functions like breathing, heart rate, and digestion. This is the bridge between the brain and the spinal cord, orchestrating the symphony of bodily functions that keep us alive.

This extraordinary architecture is not only responsible for our conscious experiences but also governs the autonomic functions that keep us ticking day in and day out. It is a testament to the complexity and wonder of the human brain, a true masterpiece of evolution.

There are a lot of particular phenomena associated with the brain's functioning.

For instance, the notion of phantom limbs is related to the sensation amputees experience, feeling their missing extremities. This intriguing phenomenon reveals the brain's unwavering commitment to its own perception of the body, even when reality tells a different story.

The McGurk Effect is a playful trick our senses play on us. Visual and auditory cues blend together, sometimes creating a whole new perception.

Imagine a rubber hand and your brain questioning its own allegiance as strokes sync with your hidden hand and a rubber surrogate. The boundary between "self" and "other" blurs, prompting fascinating questions about our own body and brain.

In the realm of neuroscience and psychology, Oliver Sacks' "The Man Who Mistook His Wife for a Hat" stands as a treasure trove of captivating clinical anecdotes that illuminate the complexities of the human mind. The book is a list of anecdotes that offer profound insights into various neurological conditions, challenging our understanding of perception, identity, and consciousness.

For instance, in the chapter "The Man Who Mistook His Wife for a Hat", Sacks' iconic tale opens our eyes to the intricacies of visual agnosia, where a man's perception is so profoundly altered that he attempts to wear his wife's head like a hat. It's a poignant reminder of how our brains construct our reality, and how this construction can sometimes falter.

In "The Lost Mariner", he discusses a man robbed of his ability to form new memories due to severe amnesia. His life becomes a series of fragmented moments, a constant present untethered from his personal history. It raises questions about the essence of self and the role of memory in shaping our identities.

In "The President's Speech", we witness the astonishing power of the brain to rally in moments of great significance. Despite language impairment, a stroke survivor delivers a flawless and impassioned speech in the presence of the President. It underscores the brain's capacity for resilience and the profound connection between language and emotion.

In "Witty Ticcy Ray" instead, he introduces us to a man with Tourette's syndrome, where verbal tics manifest as eloquent Shakespearean quotes and witty remarks. It exemplifies how the brain's intricate circuitry can give rise to unexpected expressions of creativity and linguistic prowess.

The book by Sacks is without any doubt eye-opening on the functioning of the brain and its ability to adapt to situations.

Since our focus is on neuromorphic devices, we want however to discuss in particular the learning ability of the brain.

The neuroscience of learning is a captivating field that explores how our brains process and store information, ultimately enabling us to acquire new knowledge and skills. One of the key concepts at the heart of learning is neuroplasticity. This term refers to the brain's remarkable ability to adapt and reorganize itself based on experiences and learning. It is like the brain's way of remodeling itself to accommodate new information.

Imagine the brain as a dynamic, ever-changing landscape. When we learn something new, whether it's a language, a musical instrument, or a new skill, the brain undergoes physical changes. It forms new connections between neurons, creating pathways dedicated to this newfound knowledge. It's akin to carving out a new trail through a forest, making it easier to navigate in the future.

As we have seen, this phenomenon is known as synaptic plasticity. It is the brain's way of fine-tuning its communication system. Neurons communicate through synapses, tiny gaps where chemical signals pass from one neuron to the next. Through learning, these synapses can become stronger or more efficient. It's like optimizing the transmission lines of communication in a busy city.

But it is not just about motor skills. This process applies to all forms of learning, from memorizing facts to developing problem-solving strategies. When we engage with new information, our brain works tirelessly behind the scenes, reshaping itself to accommodate this knowledge.

Furthermore, learning is not a one-size-fits-all process. Each individual's brain has its unique way of organizing and processing information. This is why some people excel in mathematics, while others thrive in artistic endeavors. It's a testament to the incredible diversity and adaptability of the human brain.

The brain's plasticity is intimately connected to the concept of complex adaptive systems, and it exemplifies an emergent phenomenon in a remarkable way.

As we have seen, a complex adaptive system is a dynamic network of interconnected elements (or agents) that exhibit collective behavior and self-organization. These systems adapt and evolve over time in response to changes in their environment or internal states. They are characterized by feedback loops, non-linear interactions, and the capacity to generate novel and unforeseen patterns, all things that have inspired the field of cybernetics.

Now, let's relate this to the brain's plasticity. The brain is a quintessential example of a complex adaptive system. It consists of a vast network of neurons that are constantly interacting with each other. When we learn new information or acquire a new skill, the connections between neurons (synapses) change. This is a dynamic process, much like the interactions between agents in a complex adaptive system.

Consider the process of learning to play a musical instrument. Initially, the neural connections associated with finger movements, hand-eye coordination, and auditory processing are relatively weak. However, with practice, these connections become stronger and more efficient. This adaptation is akin to how agents in a complex system adjust their interactions based on feedback and experience.

Another key aspect is the concept of emergence. In complex systems, emergent phenomena are behaviors or properties that arise from the interactions of individual elements but cannot be predicted by examining those elements in isolation. In the brain, learning and memory are emergent phenomena. The ability to store and retrieve information arises from the collective behavior of neurons, and it's not directly reducible to the behavior of individual neurons.

As we have seen, ant colonies and slime molds are excellent examples of decentralized, self-organizing systems. In ant colonies, individual ants operate based on simple rules and local interactions. They communicate through chemical signals (pheromones) and respond to changes in their environment and the actions of other ants. This decentralized decision-making allows the colony to exhibit coordinated behaviors like foraging, nest building, and defense.

Similarly, slime molds are single-celled organisms that can aggregate into multicellular structures when faced with a challenge, like finding food. They communicate through chemical signals and adjust their behavior based on the concentration of these signals. This decentralized cooperation allows slime molds to form complex patterns and make efficient decisions about resource allocation.

Now, let's draw a parallel to the brain's plasticity. In both cases, we see emergent behavior arising from interactions between simple elements. Just as neurons in the brain adjust their connections based on local activity, ants and slime molds adjust their behavior based on local signals and interactions.

Furthermore, all three systems—ant colonies, slime molds, and the brain—demonstrate adaptability in response to changes in their environment. Ant colonies can dynamically shift their foraging patterns based on the availability of food sources. Slime molds can adjust their structure and behavior in response to changes in resource distribution. Similarly, the brain's plasticity allows it to adapt and rewire itself in response to learning experiences and environmental cues.

Additionally, all three systems showcase the ability to solve complex problems through collective intelligence. Ant colonies can find the shortest path to a food source using a phenomenon known as ant colony optimization. Slime molds can effectively navigate mazes to find the most efficient routes to resources. In the brain, the collective behavior of neurons enables higher-order cognitive processes like problem-solving, decision-making, and creativity.

First, let's dive again into spiking neurons. We have started this book by discussing why neurons spike and the model that describes spikes. However, we have only discussed the mathematical aspects of the neurons, and we wish now to explain a bit more how this works in the visual cortex.

In biological neurons, particularly in regions like the visual cortex, the rate of spiking is intricately connected to the stimuli they receive. Let's use the visual cortex as an example to illustrate this phenomenon.

The visual cortex is a crucial part of the brain responsible for processing visual information from the eyes. Neurons in the visual cortex are highly specialized to respond to specific features in the visual field, such as edges, colors, and shapes. Each neuron has a "receptive field," which is a specific region of the visual field that it's most sensitive to.

Each neuron in the visual cortex has a preferred stimulus that it responds most strongly to. For example, one neuron might be particularly responsive to vertical lines, while another might prefer diagonal lines. When a neuron is presented with its preferred stimulus, it tends to spike more frequently.

Neurons in the visual cortex exhibit what's known as "response tuning." This means that they have a specific range of stimuli that they're most sensitive to. For instance, a neuron tuned to vertical lines might respond well to lines that are almost perfectly vertical but less so to lines at different angles.

In the visual cortex, information is encoded through the collective activity of many neurons. Different neurons have different preferences, and their combined activity encodes complex visual information. For example, the orientation and motion of an object might be represented by the combined firing rates of multiple neurons with different response properties.

Neurons in the visual cortex exhibit adaptation, which means that their response to a stimulus can change over time. For example, if a neuron is repeatedly exposed to the same stimulus, its firing rate may decrease. This adaptation helps in coding changes in contrast and sharpness of visual stimuli.

Neurons in the visual cortex also integrate information from different receptive fields to create a holistic representation of the visual scene. This integration is crucial for processes like object recognition and spatial awareness.

Let us now explain why learning can be associated with spiking neurons. We wish to consider the phenomenon of plasticity in these networks. Synaptic plasticity refers to the ability of the strength of connections (synapses) between neurons to change over time. This is fundamental for learning and memory. In spiking neural networks, this plasticity is often influenced by the timing of spikes.

This brings us to Hebbian learning, a key concept in neural network theory. Hebbian learning is a fundamental concept in neuroscience and artificial neural networks, named after the Canadian psychologist Donald Hebb. It provides a simple yet powerful explanation of how connections between neurons in the brain can be modified to facilitate learning and memory. Hebbian learning states that "neurons that fire together, wire together." When two neurons are active at the same time and contribute to the firing of a third neuron, the connection between the first two neurons is strengthened. In other words, if neuron A consistently triggers neuron B, the synapse (connection) between them becomes more efficient.

When two neurons spike together, it's a strong indication that they're involved in a common task or activity. The brain interprets this synchronous activity as a sign that these neurons are working together to achieve a specific goal. Strengthening the connection between them enhances their ability to work in tandem, making them more effective partners in future tasks.

For example, imagine learning to recognize a specific pattern, like a particular sound or visual stimulus. When neurons responsible for processing this pattern spike together, Hebbian learning reinforces their connection. This means that in the future, when one of these neurons fires in response to the pattern, it's more likely to trigger the other neurons associated with the same pattern, facilitating accurate recognition.

This concept aligns with the brain's plasticity, which is its ability to adapt and rewire itself in response to experiences and learning. When we repeatedly engage in a particular activity or thought process, the corresponding neural pathways become more efficient and robust. This is why practice and repetition are often essential for learning and skill acquisition.

Consider the example of learning to ride a bicycle. Initially, it requires focused attention and effort to coordinate balance and movement. As you practice, the neural connections associated with biking become stronger. Eventually, the process becomes more automatic, and you can ride without consciously thinking about each movement.

Hebbian learning is also intimately connected to synaptic plasticity. Synaptic plasticity refers to the ability of the strength of connections (synapses) between neurons to change. When neurons fire together, the synapses between them become more efficient, allowing for faster and more reliable transmission of signals. This is crucial for tasks like memory formation and information processing.

Additionally, Hebbian learning plays a role in the formation of neural networks. As neurons repeatedly activate one another, they form functional circuits that work together to process information and perform tasks. This collective behavior is a cornerstone of how the brain functions.

The Hopfield-Tank model is a significant contribution to the field of neural networks and associative memory. It was developed by John Hopfield and David Tank in the early 1980s. Associative memory is a type of memory that enables the retrieval of information based on its content, rather than relying on explicit addressing or cues. It's akin to the way our brains retrieve memories. In associative memory, when a partial or degraded version of a stored pattern is presented, the system is capable of completing it, producing the entire stored pattern.

Think of a well-organized library where books are arranged by subject matter. In an associative memory system, if you provide a vague description or a fragment of a book's content, the system can retrieve the entire book for you. This is in contrast to a traditional computer memory system, where you would need to know the exact location or "address" of the book.

Associative memory is particularly useful in situations where the available information is incomplete or noisy. It allows for robust retrieval of stored patterns even when the input is not an exact match.

In artificial neural networks, models like the Hopfield network and the Hopfield-Tank model are examples of systems designed to exhibit associative memory properties. These models store patterns in the connections between neurons, allowing them to retrieve complete patterns from partial or degraded inputs.

Associative memory is a key concept in both artificial intelligence and neuroscience, as it mirrors the way human memory often functions. In our brains, we don't always need precise cues to recall a memory; sometimes, a fragment or related piece of information is enough to trigger a complete recollection.

Going back to the Hopfield and Hopfield-Tank model, these models are particularly interesting because they demonstrate how a network of simple units, or neurons, can perform complex tasks such as pattern recognition and memory retrieval.

Imagine then a network of interconnected neurons, where each neuron can be in one of two states: firing or not firing. The Hopfield-Tank model explores how these neurons can be connected in a specific way to store and retrieve patterns.

The genius of the model lies in its use of associative memory. In traditional computer memory, you retrieve information by providing an address, and the memory returns the data stored at that address. In associative memory, on the other hand, you provide a partial or degraded version of the stored pattern, and the memory retrieves the complete pattern that it associates with it. This is similar to how human memory often works - a hint or cue can trigger the recall of related information.

The Hopfield-Tank model is based on the idea that connections between neurons are symmetric. This means that if neuron A is connected to neuron B, then neuron B is also connected to neuron A with the same strength. These connections are weighted, meaning some connections are stronger than others. This architecture allows for the formation of stable attractor states, which are patterns that the network naturally tends towards.

The model demonstrated how, even with a relatively simple set of rules and connections, a network of neurons could be trained to reliably retrieve stored patterns from partial or noisy inputs. This had implications for areas like content-addressable memory and pattern recognition in artificial intelligence.

Hebbian learning and the Hopfield-Tank model can also be expressed mathematically. The change in connectivity (synaptic weight) between two neurons changes according to the Hebbian (product law), which can be mathematically expressed as:

$$\Delta W_{ij} = \eta x_i x_j$$

where ΔW_{ij} denotes the change in the synaptic weight between neurons i and j, η represents the learning rate, and x_i and x_j signify the activations of the respective neurons.

Building upon this, the Hopfield-Tank model emerges as a profound extension. It embodies the idea of associative memory, where patterns can be recalled from partial or degraded inputs. In this model, a network of binary neurons, each with an "on" or "off" state, is connected by symmetric synaptic weights. The state evolution is governed by a discrete-time dynamics:

$$x_i(t + 1) = \text{sgn}\left(\sum_{j=1}^{N} W_{ij} x_j(t) - \theta_i\right)$$

where $x_i(t)$ is the state of neuron i at time t, W_{ij} represents the synaptic weight between neurons i and j, and θ_i is the threshold for neuron i.

Through this iterative process, the network converges to stable states, which correspond to stored patterns. This process beautifully encapsulates the concept of memory recall based on partial cues.

The mathematical elegance of the Hopfield-Tank model lies in its ability to encode memories as fixed points in a high-dimensional space. These fixed points serve as attractor states, capturing the essence of associative memory.

Another interesting aspect of the Hopfield-Tank model is that it can be derived from circuits of resistors and capacitors. Let's explain this connection.

In electrical engineering, RC circuits are commonplace components used for various purposes, including signal processing and filtering. These circuits consist of a resistor (R) and a capacitor (C) connected in series or parallel. The capacitor stores and releases electrical energy over time, while the resistor regulates the flow of current. As we have seen, RC networks modeled the spiking neurons of the giant axon.

Now, consider the Hopfield-Tank model, which is a recurrent neural network known for its associative memory properties. It is composed of binary neurons, each capable of being in an "on" or "off" state (as an approximation of discrete spikes). The connections between neurons are symmetric, and the state evolution is determined by a discrete-time dynamics equation.

The intriguing connection arises from the mathematical similarity between the dynamics of the Hopfield-Tank model and the behavior of RC circuits. Specifically, the differential equations governing the voltage across a capacitor in an RC circuit and the state evolution equation in the Hopfield-Tank model share a similar form.

In both cases, there is a temporal component (time derivative in the RC circuit, discrete time steps in the neural network) and a weighted sum of inputs (across the resistor in the RC circuit, and weighted inputs in the neural network). This parallel underscores the elegance of how concepts from electrical engineering and neural network theory converge.

Furthermore, this analogy allows us to gain fresh insights into the behavior of neural networks. For instance, just as an RC circuit can store and release electrical energy, the Hopfield-Tank model can store and retrieve patterns of neural activity. This correspondence between electrical circuits and neural networks offers a unique perspective on how information processing can be understood and applied across different domains. Now, it is important to note that there is a way to explain why the dynamics of the neuronal states in the Hopfield-Tank model lead to a state retrieval in the long-time regime.

In the continuous Hopfield-Tank model, the dynamics of the system are governed by differential equations. These equations describe how the state of each neuron changes over time as a function of the current state and the synaptic connections. The continuous-time dynamics can be represented as:

$$\dot{x}_i = -x_i + \sum_{j=1}^{N} W_{ij} x_j - \theta_i$$

Here, x_i represents the state of neuron i, W_{ij} denotes the synaptic weight between neurons i and j, and θ_i is the threshold for neuron i.

To derive a Lyapunov function from this model, one typically seeks a scalar function $V(x)$ that satisfies two conditions:

1. **Positivity**: $V(x) > 0$ for all x in the state space, except at the equilibrium points.

2. **Decrescence**: $\dot{V}(x) \leq 0$ for all x in the state space, except at the equilibrium points where $\dot{V}(x) = 0$.

The Lyapunov function essentially serves as a measure of the system's energy or a "potential" that characterizes its stability, and thus also the convergence.

Now, the key insight in this context is that, under certain conditions on the synaptic weights and thresholds, it is possible to find a Lyapunov function that demonstrates that the continuous Hopfield-Tank model converges to stable states or attractors.

This Lyapunov function essentially quantifies how the system's energy decreases over time, indicating convergence to a stable state. It provides a mathematical guarantee of the model's stability properties.

It's worth noting that the specific form of the Lyapunov function and the conditions for its existence and uniqueness depend on the details of the Hopfield-Tank model, including the synaptic weights, thresholds, and network architecture.

In summary, the derivation of a Lyapunov function from the continuous Hopfield-Tank model is a powerful technique for analyzing the stability and convergence properties of the neural network. It provides a rigorous mathematical framework for understanding the behavior of the system over time.

The Hopfield-Tank model stands as a pivotal milestone in the realms of neuroscience and artificial intelligence. Its historical significance lies in its ambitious endeavor to unravel the intricacies of memory retrieval.

At the time of its inception in the 1980s, the model represented a pioneering leap forward. It embodied the resurgence of interest in neural networks, presenting a mathematical framework that showcased how networks of rudimentary computational units, or neurons, could collectively perform intricate cognitive tasks such as memory retrieval.

One of the model's groundbreaking features was its introduction of content-addressable memory. This revolutionary concept enabled the retrieval of stored information based on partial or degraded cues, deviating from the conventional exact addressing used in computer memory.

Furthermore, the Hopfield-Tank model illuminated the principle of associative memory. It demonstrated the network's capability to recall patterns from partially specified or noisy inputs, mirroring how human memory functions. This departure from earlier models, which required exact matches, brought the model closer to biological memory processes.

The concept of attractor dynamics, introduced by the model, further advanced our understanding of memory retrieval. The network exhibited a tendency to settle into stable states, or attractors, which represented stored memories. This property elucidated how memories could be robustly retrieved even in the presence of noisy or partial cues.

Inspired by the biological brain, the model echoed the notion that memory storage and retrieval in the brain are distributed and based on patterns of activity across interconnected neurons.

Beyond its contributions to memory retrieval, the Hopfield-Tank model demonstrated its broader applications. It revealed connections to optimization problems, hinting at the potential for similar principles to be applied in solving a wide range of computational tasks.

The model's enduring legacy lies in its influence on subsequent research in neural networks and associative memory. It laid the foundation for modern developments in machine learning and artificial intelligence.

In essence, the Hopfield-Tank model marked a pivotal step towards bridging the gap between neural network theory and the computational principles underpinning memory retrieval.

In conclusion, the exploration of the brain and its computational models has unveiled remarkable parallels between biological systems and artificial constructs. The Hopfield model, with its associative memory properties, provided a pioneering glimpse into the potential of neural networks to mimic the brain's ability to retrieve information based on content, rather than rigid addresses.

This model, inspired by the intricate interplay of neurons in the brain, has paved the way for the development of sophisticated artificial neural networks. These networks, designed to process information in a distributed and parallel manner, have become foundational in modern artificial intelligence.

A crucial aspect that links these artificial networks with their biological counterparts is the concept of plasticity. In both cases, whether in the synapses of the brain or the weights in a neural network, adaptability is key. Plasticity enables learning, allowing the system to adjust its connections based on experience and input. This adaptive capability is fundamental to the brain's remarkable ability to learn and adapt to new information.

As we transition to the next chapter on modern artificial intelligence, we embark on a journey through the landscape of modern artificial intelligence toward future neuromorphic devices. Armed with insights gleaned from the study of the brain and its computational models, we'll explore how contemporary AI approaches harness the power of neural networks, deep learning, and advanced algorithms to tackle complex tasks and unlock new frontiers in artificial intelligence research and applications. Through this evolution, we will witness the continued convergence of biological inspiration and cutting-edge technology, leading to unprecedented advances in the field of artificial intelligence.

4 Modern AI and computers

> Early AI was mainly based on logic. You're trying to make computers that reason like people. The second route is from biology: You're trying to make computers that can perceive and act and adapt like animals.

Geoffrey Hinton

From the seminal work of Warren McCulloch and Walter Pitts in the 1940s which we discussed earlier, which laid the foundation for artificial neurons and basic computational units, the field of neural networks has undergone a remarkable evolution.

McCulloch and Pitts introduced the simplified model of biological neurons, demonstrating that even these simple binary units could perform complex logical operations. This marked the birth of artificial neural networks and set the stage for further exploration.

In the decades following, researchers extended their work, culminating in the development of the perceptron by Frank Rosenblatt in the late 1950s. The perceptron, a single-layer neural network, was capable of linearly separating classes in simple tasks, sparking optimism about the potential of neural networks.

However, the perceptron's limitations in handling non-linearly separable tasks led to a period known as the "AI winter," where neural networks fell out of favor in favor of rule-based systems and other approaches.

The resurgence of neural networks came with the introduction of backpropagation, a powerful algorithm for training multi-layer neural networks, in the 1970s. This breakthrough allowed networks with multiple hidden layers (known as multi-layer perceptrons or MLPs) to learn complex, non-linear relationships in data.

In the late 1980s and 1990s, researchers like Geoffrey Hinton and Yann LeCun made significant contributions, further advancing the field. Convolutional neural networks (CNNs) emerged, revolutionizing image processing tasks. Hinton's work on deep belief networks and LeCun's pioneering work on convolutional neural networks laid the groundwork for modern deep learning.

The 2010s saw an explosion in deep learning research, fueled by the availability of large datasets and powerful computational resources. Breakthroughs in architectures (e.g., recurrent neural networks, long short-term memory networks) and techniques (e.g., transfer learning, generative adversarial networks) propelled deep learning to the forefront of artificial intelligence.

In recent years, the field of artificial intelligence (AI) has experienced a transformative revolution, reshaping how machines perceive, learn, and interact with the world around them. While at the forefront of this revolution are deep neural networks, powerful computational models inspired by the structure and function of the human brain have been derived.

Deep neural networks, often referred to simply as deep learning, represent a paradigm shift in AI. Unlike earlier approaches that relied on handcrafted features and explicit programming, deep learning leverages hierarchical layers of artificial neurons to automatically learn abstract features from data. This enables the system to discern complex patterns and make high-level abstractions, similar to how our brains process information.

The term "deep" arises from the multiple layers that compose these networks. Each layer refines the representation of the input, passing it forward through the network. This deep architecture allows for the extraction of increasingly sophisticated features, enabling the network to understand intricate relationships within the data.

The success of deep neural networks can be attributed to their ability to tackle tasks that were once considered insurmountable for machines. From image and speech recognition to natural language processing and even playing complex games, deep learning has demonstrated unprecedented levels of performance and accuracy.

What sets deep neural networks apart is their capacity for end-to-end learning. Rather than relying on explicit rules and human-designed features, these networks learn directly from raw data, automatically discovering the most relevant features and representations. This ability to learn from data has made deep learning particularly effective in domains with large, complex datasets.

The widespread adoption of deep neural networks has led to groundbreaking advancements in numerous fields. In healthcare, they're used for medical image analysis and diagnosis. In autonomous vehicles, they power perception systems that enable safe navigation. In finance, they enhance fraud detection and risk assessment. These applications represent just a fraction of the transformative impact deep learning has had on a wide range of industries.

However, before we discuss deep neural networks, let us discuss the simplest model of neural network.

Single-layer neural networks, also known as perceptrons, are the simplest form of artificial neural networks. They consist of only one layer of artificial neurons (perceptrons) that take inputs, apply weights to them, and produce an output based on a specified activation function.

The perceptron algorithm, which was developed by Frank Rosenblatt in the late 1950s, made single-layer neural networks possible. The key components of the perceptron algorithm are as follows:

Inputs and Weights: Each input is assigned a weight, which determines its relative importance. The inputs are multiplied by their respective weights, and these weighted inputs are summed up.

Activation Function: The weighted sum is then passed through an activation function. In the original perceptron, this function was a simple step function. If the sum exceeds a certain threshold, the output is set to 1; otherwise, it's set to 0.

Output: The output of the perceptron is the result of the activation function.

The perceptron algorithm is capable of learning simple linear decision boundaries, making it suitable for tasks that are linearly separable (e.g., binary classification tasks where a straight line can separate the two classes).

However, perceptrons have limitations. They cannot learn more complex patterns that require nonlinear decision boundaries. This led to the development of multi-layer neural networks, also known as multi-layer perceptrons (MLPs), which can learn much more intricate relationships in the data.

MLPs incorporate multiple layers of neurons, allowing for the creation of highly nonlinear models. They utilize more sophisticated activation functions (such as sigmoid or ReLU) that can introduce nonlinearity into the network's computations. We have seen these functions before, and thus it should not come as a surprise.

The backpropagation algorithm, which was independently developed by multiple researchers in the 1970s, is a crucial advancement that made training multi-layer neural networks feasible. Backpropagation is a gradient-based optimization technique that allows the network to adjust its weights in response to errors in its predictions.

Single-layer neural networks (perceptrons) rely on the perceptron algorithm, which involves assigning weights to inputs and applying an activation function. However, their capability is limited to linearly separable tasks. Multi-layer neural networks (MLPs), on the other hand, overcome this limitation and can learn highly complex patterns. The development of the backpropagation algorithm was instrumental in training these deeper networks.

Artificial neural networks possess a remarkable property that sets them apart in the realm of machine learning: they are considered universal approximators. This means that, given a sufficiently large number of neurons in their hidden layers, they have the extraordinary ability to approximate any continuous function with arbitrary precision.

This capability arises from the intricate structure of neural networks. Comprising multiple layers of interconnected neurons, each layer refines the representation of the input data, gradually extracting more abstract features. This hierarchical approach allows neural networks to capture complex relationships in the data, mimicking the way our brains process information.

Crucially, the inclusion of nonlinear activation functions is fundamental to the network's expressive power. These functions introduce nonlinearity into the computations performed by the neurons. Without this nonlinearity, the network would be limited to performing linear transformations, severely constraining its capacity to model complex relationships.

Consider a neural network as a series of compositions of simpler functions. Each neuron conducts a weighted sum followed by a nonlinear activation function. These compositions, linked through the layers, empower the network to approximate highly complex, nonlinear functions.

The Universal Approximation Theorem, established by mathematicians like George Cybenko, formally confirms this extraordinary capability. It asserts that a single-hidden-layer neural network, employing a nonconstant, bounded, and continuous activation function, can approximate any continuous function on a compact domain to any desired level of accuracy.

Yet, achieving this theoretical capability in practice can be demanding. The number of neurons required for accurate approximation can be exceedingly large, making training computationally intensive and susceptible to overfitting. Furthermore, the choice of architecture, activation functions, and training algorithms significantly influences the network's ability to approximate complex functions.

In essence, the universal approximation property of neural networks underscores their versatility and power in modeling a wide range of real-world phenomena of neural networks. Going back to the perceptron, mathematically it is very similar to the neural networks we had discussed before.

The perceptron is a fundamental building block of neural networks and is often used as a binary linear classifier (which we will explain shortly) that makes decisions based on a weighted sum of inputs, followed by a threshold function. Mathematically, the output y of a single perceptron can be represented as:

$$y = f\left(\sum_{i=1}^{n} w_i x_i + b\right)$$

Here, x_i represents the i-th input, w_i denotes the weight associated with the i-th input, b is the bias term, and f is the activation function.

The perceptron makes a decision based on whether the weighted sum plus bias is greater than or equal to zero. If the condition is met, the output is set to 1 (or "on"); otherwise, it's set to 0 (or "off"). This behavior is akin to a binary switch, making the perceptron capable of linearly separating two classes.

In a graphical sense, this operation can be visualized as defining a hyperplane in the input space that partitions it into two regions, each corresponding to one of the output classes.

Learning in a perceptron involves adjusting the weights and bias terms based on the training data. The weights are updated using a learning algorithm, such as the perceptron learning rule, which aims to minimize the error in classification.

It's important to note that perceptrons can only learn linearly separable functions. This means they are limited to tasks where a straight line (or hyperplane in higher dimensions) can cleanly separate the classes.

While perceptrons on their own have limitations, they form the foundation for more complex neural network architectures, including multi-layer perceptrons (MLPs) that can learn highly nonlinear relationships in data. These extensions, incorporating multiple layers and sophisticated activation functions, significantly broaden the capabilities of neural networks.

A classifier is a computational model or algorithm that assigns a label or category to a given input based on its characteristics or features. In the context of machine learning and artificial intelligence, a classifier is typically trained on a dataset with labeled examples (input-output pairs) in order to learn patterns and make predictions or decisions about unseen or future data.

For instance, in a binary classification task, a classifier aims to divide input data into one of two classes or categories. This could be as simple as distinguishing between "spam" and "non-spam" emails based on features like the content, sender, and subject line.

In multi-class classification, the goal is to assign data into one of several categories. For example, a classifier might be trained to recognize handwritten digits (0-9) based on pixel values.

Classifiers can take various forms, including linear classifiers (like the perceptron or support vector machines), decision trees, random forests, neural networks, and more. Each type of classifier has its own strengths, limitations, and applications depending on the nature of the data and the problem at hand.

The extension of the perceptron model to multi-layers is called deep neural network (DNN). To put it in perspective, think about a single-layer neural network as a basic tool that can handle straightforward tasks. It's like a simple calculator that can perform basic operations. It is useful, but it has its limitations. It struggles when faced with more intricate problems, much like a calculator would when trying to solve a complex algebraic equation.

Deep neural networks, on the other hand, take this concept to a new level. They are a series of interconnected calculators, each one building upon the results of the previous one. This allows for the gradual discovery of intricate patterns and relationships within the data.

Picture it as a team of experts working together, where each member specializes in a particular aspect of a problem. They collaborate, with each one refining the information and passing it along until they collectively arrive at a comprehensive understanding.

The power however lies in the layers. These "hidden" layers are where the network learns to extract meaningful features from the input data. It's like peeling back the layers of an onion to reveal its core. Each layer refines the representation of the data, gradually transforming it into a form that is better suited for making accurate predictions or classifications.

This hierarchical approach to feature extraction enables DNNs to tackle incredibly complex tasks. They're adept at recognizing intricate patterns in images, understanding the nuance in language, and even making sophisticated decisions in tasks like autonomous driving and classifiers.

Classifiers find extensive use in a wide range of fields including computer vision, natural language processing, medical diagnosis, finance, and many others. They are integral to tasks like object recognition, sentiment analysis, fraud detection, and disease diagnosis, among others.

Now, there are many methods to teach machines how to perform tasks. Reinforcement learning (RL) represents a distinctive paradigm in machine learning that revolves around the concept of learning through interaction with an environment. Unlike supervised learning, where a model is trained on labeled data, or unsupervised learning, which focuses on uncovering hidden patterns, reinforcement learning is centered on decision-making and sequential actions.

In reinforcement learning, an agent interacts with an environment in discrete time steps. At each step, the agent receives a state, takes an action, and receives a reward signal from the environment. The goal of the agent is to learn a policy—a strategy for selecting actions based on states—that maximizes the cumulative reward over time.

The distinctive feature of reinforcement learning is its focus on delayed rewards. This means that the consequences of an action may not be immediately apparent, making it a challenging but vital area of study, particularly in dynamic and uncertain environments.

Now, in comparison to deep neural networks (DNNs), which have gained immense popularity in machine learning, there are notable differences. While deep neural networks are versatile function approximators capable of learning complex mappings between inputs and outputs, they are typically used in supervised and unsupervised learning scenarios.

Reinforcement learning, on the other hand, deals with sequential decision-making problems where the agent interacts with an environment and receives feedback in the form of rewards. It necessitates the agent to learn a policy that balances exploration (trying new actions) and exploitation (choosing actions based on past knowledge) to maximize long-term rewards.

However, it's important to note that reinforcement learning and deep neural networks are not mutually exclusive. Deep learning techniques, especially deep neural networks, have been successfully integrated into reinforcement learning algorithms, giving rise to what is known as deep reinforcement learning (DRL). Deep reinforcement learning leverages the power of DNNs to approximate complex value functions or policies, enabling agents to tackle high-dimensional state spaces and achieve impressive results in tasks ranging from playing video games to controlling robots.

Let us briefly explain the mathematical framework. The agent's goal (our computer or robot) in reinforcement learning is to learn a policy π that maps states to actions, aiming to maximize the cumulative sum of rewards, known as the return. This return is defined as:

$$G_t = r_{t+1} + \gamma r_{t+2} + \gamma^2 r_{t+3} + \ldots = \sum_{k=0}^{\infty} \gamma^k r_{t+k+1}$$

where γ is the discount factor ($0 \leq \gamma \leq 1$), which determines the present value of future rewards. A lower γ values the agent's preference for immediate rewards, while a higher γ values long-term rewards.

The agent's objective is to find the policy π that maximizes the expected return:

$$J(\pi) = E_\pi \left[\sum_{k=0}^{\infty} \gamma^k r_{t+k+1} \right]$$

where E represents the average with respect to the policy.

This process of finding the optimal policy is known as the RL problem. There are various algorithms to tackle this, including Value Iteration, Policy Iteration, Monte Carlo methods, Temporal Difference learning, and more. These algorithms aim to estimate the value function $V(s)$, which measures the expected return from a given state s following a policy π, or the action-value function $Q(s, a)$, which measures the expected return from taking action a in state s following policy π.

Exploration-exploitation trade-off is a critical aspect in RL. The agent must balance exploring new actions to discover potentially better strategies versus exploiting known actions that have previously yielded high rewards.

Reinforcement learning has found wide applications in areas such as robotics, game-playing, recommendation systems, and autonomous agents. It has led to remarkable achievements, from training agents to playing complex video games to controlling real-world systems like self-driving cars and robotic arms.

Current computer architectures face several limitations when it comes to efficiently processing neural networks, particularly deep learning models.

Deep learning models, especially large ones, require a significant amount of data to be shuttled between the processor and memory. Traditional computer architectures struggle to keep up with the high demands for memory bandwidth, leading to performance bottlenecks.

Neural networks, especially deep ones, are highly parallelizable. They perform numerous computations simultaneously, making them well-suited for parallel processing architectures. However, conventional CPUs have limited parallel processing capabilities compared to specialized hardware like GPUs or TPUs.

Deep learning models often demand substantial computational resources, which can lead to high power consumption. This is a significant concern for applications like mobile devices or edge computing, where power efficiency is crucial.

For real-time applications like autonomous vehicles or robotics, low latency in processing is critical. Traditional architectures may introduce delays, impacting the responsiveness of the system.

Deep learning models often require high precision in computations to maintain accuracy. However, many traditional architectures work with 32 or 64-bit precision, which can be overkill for neural network computations. More specialized hardware can perform computations at lower precision, which saves processing time and energy.

As models grow in size and complexity, scaling them to handle larger datasets or more parameters becomes a challenge for conventional architectures. Specialized hardware, like distributed computing systems or large-scale GPU clusters, are often required.

Large models, particularly those with billions of parameters, may not fit entirely within the memory of a single machine. This necessitates techniques like model parallelism or distributed training, which can be complex to implement.

General-purpose CPUs are designed to handle a wide range of tasks, but they may not be optimized for the specific computations required by neural networks. Specialized hardware, like GPUs and TPUs, are more tailored for these tasks but may lack the flexibility to handle other types of computations efficiently.

Some neural network architectures and algorithms may not be well-suited to conventional computing architectures. For instance, recurrent neural networks (RNNs) can be computationally intensive, and traditional architectures may struggle to process them efficiently.

Addressing these limitations requires the development of specialized hardware and architectures designed specifically for neural network computations. This has led to the rise of GPUs, TPUs, and other application-specific integrated circuits (ASICs) optimized for deep learning tasks. Additionally, advancements in software frameworks and algorithms are also crucial in maximizing the efficiency of current computer architectures for neural networks.

Picture a room full of workers, each tasked with performing a specific calculation. Traditional central processing units (CPUs) would approach this task by having one worker meticulously complete a task before moving on to the next. It's a sequential process, much like how a cashier at a grocery store handles one customer at a time.

Now, imagine a different scenario. Instead of one worker, you have a massive team, all capable of working simultaneously. This team operates in perfect harmony, swiftly processing a multitude of tasks in unison. This is akin to the parallel processing prowess of a Graphics Processing Unit (GPU).

In the realm of neural networks, especially during the training phase, a critical operation arises: matrix multiplication. This operation is akin to the intricate choreography of a dance. Data is arranged in matrices, with each row representing a feature and each column representing a data point. These matrices flow through the neural network, encountering weight matrices along the way.

These weight matrices hold the key to learning. They contain the parameters that the network adapts through training. To compute the activations of the next layer, these matrices must be multiplied together. It's a fundamental step in both the forward and backward passes of training.

Now, imagine executing this operation for each layer of a deep neural network. The number of computations involved is staggering. This is where the GPU truly shines. Its architecture is meticulously crafted to excel at precisely this kind of calculation, handling a multitude of them simultaneously.

The result is a remarkable acceleration of the training process. What might take hours or even days on a CPU can be accomplished in a fraction of the time with a GPU. This speed is of paramount importance, especially in the fast-paced world of research and industry.

In fact, the deep learning community has recognized the significance of GPUs and developed specialized libraries and frameworks that harness their power. These tools are finely tuned to make the most of GPU hardware, ensuring that neural network computations are executed with maximum efficiency.

As neural networks continue to grow in complexity and size, the demand for the parallel processing capabilities of GPUs becomes even more apparent. They are not just tools; they are the driving force behind the development of increasingly sophisticated and accurate neural network models, propelling advancements in fields ranging from computer vision to natural language processing and beyond.

Another important piece of hardware is the Application-Specific Integrated Circuit, or ASIC, which is purpose-built for a specific task, like a precision instrument finely tuned to excel at one job. This is the essence of an ASIC. In the realm of neural networks, where efficiency and speed are paramount, ASICs step in as the specialists.

Think of a neural network as a complex orchestra, with each neuron playing its part in a grand symphony of computations. Training and running these networks involve a multitude of operations, many of which involve matrix multiplications and other specialized computations. This is where ASICs truly shine.

Unlike a general-purpose processor, which must juggle a wide array of tasks, an ASIC is custom-crafted. It's designed from the ground up to handle the specific computations required for neural networks. This means that every transistor etched onto its surface is dedicated to the task at hand, resulting in unparalleled efficiency.

Consider the power demands of neural network computations. ASICs, being meticulously optimized, require fewer transistors and consume less power compared to their general-purpose counterparts. This is a game-changer for applications where power efficiency is not just a preference, but a critical requirement.

When it comes to latency and throughput, ASICs deliver with unmatched precision. They're engineered to process data swiftly and efficiently, crucial for real-time applications like autonomous vehicles or live video processing.

Take, for instance, the realm of cryptocurrency mining. ASIC miners are purpose-built to handle the intricate calculations required for blockchain transactions. Their efficiency in this specific task far surpasses that of general-purpose processors.

However, there's a trade-off. The development of ASICs is no small feat. It involves substantial upfront costs, from chip design to production. This makes them most feasible for applications with high production volumes or those where performance is absolutely paramount.

Once an ASIC is designed and manufactured, it's dedicated to its specific function. It's not a versatile multitool, but a finely honed instrument for a particular task. This inflexibility can be a limitation in scenarios where adaptability is crucial.

While ASICs and GPUs are powerful tools for neural networks, they do come with a notable limitation: the finite precision of Complementary Metal-Oxide-Semiconductor (CMOS) hardware.

Picture a landscape with varying degrees of detail. Now, imagine trying to represent this landscape with a limited palette of colors. The finer nuances might get lost, and some features could be oversimplified or even entirely missed. This is akin to the challenge posed by finite precision in CMOS hardware for neural networks.

In the world of deep learning, many computations rely on high precision, particularly when dealing with small gradients or very large or very small numbers. However, standard CMOS hardware is bound by limitations in terms of the precision it can handle. It operates with a fixed number of bits to represent data, and when the numbers involved become extremely small or large, precision can be compromised.

This limitation can lead to a phenomenon known as "quantization error." It's like trying to measure the exact weight of an object with a scale that only displays whole numbers. The result might be close, but it's not exact. In neural networks, this can manifest as a loss of accuracy in computations, particularly in deep networks or those dealing with highly sensitive data.

Additionally, as neural networks become increasingly sophisticated and require higher precision for tasks like natural language processing or medical image analysis, the constraints of finite precision become more pronounced. Some applications may demand precision levels that are simply beyond the capability of standard CMOS hardware.

This is where researchers and engineers are exploring innovative solutions, such as specialized hardware architectures or techniques like quantization-aware training. These approaches aim to mitigate the impact of finite precision, ensuring that neural networks can perform with the required accuracy and reliability.

The previous analogy on finite precision palettes provides insight into why storing analog information is crucial in hardware tailored for neural networks and machine learning.

In the realm of deep learning, precision matters. Some computations hinge on extremely small differentials or rely on very large or very small values. In these situations, maintaining high precision is imperative. However, standard digital hardware, which operates on discrete values, can occasionally fall short. It's akin to trying to measure an object's weight with a scale that only provides whole numbers. The result may be close, but it's not entirely accurate.

This is where analog information storage steps in as a game-changer. Analog hardware operates on a continuous scale, much like a spectrum of colors, allowing for a finer level of detail. It's like having a painter's palette with an array of shades to choose from, enabling a more nuanced representation of data.

Analog storage allows for the representation of values in a continuous range, mitigating the effects of quantization errors that can arise in purely digital systems. It's akin to providing a canvas with infinite color possibilities, ensuring that even the most subtle gradients and intricate details are preserved.

Furthermore, analog hardware is particularly adept at handling signals that vary smoothly over time. This is invaluable in tasks like signal processing, where the precise representation of dynamic data is essential. For example, in applications like audio processing or image manipulation, analog storage can capture the nuances that may be missed in a purely digital framework.

By embracing analog information storage, we unlock the potential for more precise, detailed, and nuanced computations in neural networks and machine learning systems. This is especially vital as these technologies continue to advance, tackling increasingly complex tasks that demand a high level of accuracy and sensitivity.

5 Technologies behind neuromorphic devices

> It's easy to have a complicated idea. It's very hard to have a simple idea.

Carver Mead

Carver Mead, a visionary figure in the field of neuromorphic computing, has championed a paradigm shift in the way we conceive and build computing systems. Drawing inspiration from the intricate workings of the human brain, Mead envisioned a computing approach that emulates the efficiency, adaptability, and energy-conscious nature of neural networks.

At the heart of Mead's vision is the recognition that conventional digital computing, while immensely powerful, can be inefficient when it comes to tasks that closely resemble the parallel, distributed processing seen in biological systems. In contrast, the brain, composed of billions of neurons, excels at tasks like pattern recognition, sensory processing, and learning through synaptic plasticity.

Mead's neuromorphic computing departs from the conventional von Neumann architecture, which separates processing and memory, and instead seeks to integrate computation and memory in a manner that mirrors the brain's neural networks. By doing so, neuromorphic systems aim to process information in a more brain-like fashion, allowing for tasks that demand high parallelism, low energy consumption, and real-time responsiveness.

After understanding the von Neumann architecture, it's important to explore the paradigm of In-Memory Computing, which represents a departure from the traditional computing model.

In the von Neumann architecture, there is a clear separation between the processing unit (CPU) and the memory. This means that data is shuttled back and forth between the memory and the processor, incurring a significant amount of time and energy overhead.

In contrast, In-Memory Computing, as the name suggests, performs computations directly within the memory itself. This is achieved by embedding processing elements, such as computational units or accelerators, directly into the memory cells or within the memory hierarchy. This allows for parallel processing of data at the location where it is stored, eliminating the need for continuous data movement between memory and processor. This is related to Mead's vision for analog computers. Mead's approach advocates for the utilization of analog circuits, which can process information in a continuous, gradient-like manner, akin to the way neurons communicate through graded potentials. This departure from the discrete nature of digital logic allows for more natural and energy-efficient computations, particularly in tasks that require nuanced, real-valued representations. One of Mead's pioneering contributions is the development of silicon neurons and synapses, which serve as building blocks for these neuromorphic systems. These artificial neurons mimic the behavior of their biological counterparts, and when interconnected, form networks capable of complex computations.

Carver Mead's ideas have ignited a new era in computing, where the focus is not solely on raw computational power, but on efficiency, adaptability, and the ability to tackle tasks that were once the domain of biological systems. His work has inspired a vibrant community of researchers and engineers dedicated to realizing the full potential of neuromorphic computing, with the hope of creating machines that truly learn, perceive, and interact with the world in ways that rival and, in some aspects, even surpass human capabilities. This vision was, in the early 90's, incorporated into Very Large Scale Integration (VLSI), which were at the forefront of a transformative shift in the field of electronic design. VLSI started earlier, however, spanning the late 1960s to the 1970s, saw the emergence of a revolutionary approach to creating integrated circuits. VLSI technology allowed for the integration of thousands, and eventually millions, of transistors onto a single silicon chip.

Mead's contributions to VLSI design were groundbreaking, and his work laid the foundation for modern semiconductor technology. Mead's pioneering insights into VLSI design principles were instrumental in enabling the miniaturization and integration of electronic components. He emphasized the importance of designing circuits at a higher level of abstraction, advocating for the use of silicon compilers that could automatically generate layouts from high-level descriptions.

Furthermore, Mead was a proponent of the analog approach to VLSI, which focused on harnessing the continuous properties of electronic signals. This perspective stood in contrast to the prevalent digital design methodologies of the time. Mead's advocacy for analog VLSI paved the way for more energy-efficient and versatile electronic systems.

In collaboration with Lynn Conway, Carver Mead co-authored the influential textbook "Introduction to VLSI Systems," which became a seminal work in the field. The book not only provided a comprehensive overview of VLSI design principles but also introduced innovative concepts that would shape the future of electronic engineering.

Mead's work in VLSI design not only revolutionized the semiconductor industry but also had a profound impact on various fields, including computer architecture, artificial intelligence, and neuromorphic computing. His visionary approach to circuit design continues to inspire generations of engineers and researchers, leaving an indelible mark on the landscape of modern electronics.

The past two decades have witnessed a remarkable evolution in the field of neuromorphic computing, marking a transition from the dominance of Very Large Scale Integration (VLSI) to a diverse array of innovative neuromorphic systems.

Initially, VLSI technology reigned supreme, enabling the integration of millions of transistors on a single chip. This heralded a new era of electronic design, facilitating the development of powerful digital systems. However, as computational demands expanded to encompass tasks requiring parallelism, adaptability, and energy efficiency akin to biological neural networks, a paradigm shift began to emerge.

Carver Mead's pioneering work in the 1970s laid the groundwork for neuromorphic computing by emphasizing the importance of analog circuitry and higher-level design abstraction. Building on Mead's insights, researchers and engineers began exploring alternative approaches to computing that leveraged the principles of neuroscience and analog processing.

One notable departure from traditional VLSI was the advent of memristors, resistors with memory, which promised to mimic synaptic behavior in artificial neural networks. This breakthrough introduced a novel hardware element that played a pivotal role in neuromorphic system development.

Spiking Neural Networks (SNNs) also emerged as a significant departure from traditional von Neumann architecture. These networks modeled information processing based on the spiking behavior of neurons, aligning more closely with the biological workings of the brain. This shift allowed for more energy-efficient and biologically plausible computations.

Additionally, the exploration of novel materials and technologies, such as phase-change memory (PCM) and optoelectronic components, further expanded the toolkit for neuromorphic system designers. These elements introduced capabilities like non-volatility, rapid switching, and photon-based information processing, opening new avenues for efficient and high-speed computations.

In recent years, the field of neuromorphic computing has diversified even further, encompassing biohybrid systems, quantum neuromorphic computing, and synthetic neurobiology. These approaches combine elements of biology, quantum mechanics, and synthetic biology to create hybrid computational platforms with unprecedented capabilities.

This historical transition from VLSI to a rich tapestry of neuromorphic systems reflects a concerted effort to bridge the gap between conventional computing and the remarkable processing capabilities of the human brain. By drawing inspiration from biology and embracing diverse hardware platforms, researchers aim to create a new generation of intelligent machines capable of tasks that were once the realm of science fiction.

At its essence, neuromorphic computing seeks to replicate the parallel and distributed nature of neural processing. Unlike the von Neumann architecture that segregates computation and memory, neuromorphic systems leverage specialized hardware to integrate these functions. This deviation holds the potential to surmount the bottlenecks inherent in traditional computing paradigms, particularly for tasks that demand massive parallelism and low power consumption.

The hallmark of neuromorphic hardware lies in its capacity for analog emulation of neurons and synapses. Employing analog circuits and components, these systems approximate the continuous nature of biological processes, allowing for nuanced and efficient information processing. This attribute enables applications such as pattern recognition, learning, and inference to be executed with unprecedented efficiency, propelling advancements in domains spanning computer vision, natural language processing, and beyond.

A distinguishing feature of neuromorphic computing is its intrinsic plasticity, a facet that mirrors the adaptability of biological neural networks. This plasticity holds immense promise for tasks necessitating continual learning and adaptation to evolving data streams. It introduces a dynamic element into computations, making neuromorphic systems adept at tasks ranging from real-time decision-making to dynamic pattern recognition.

Consider a task that requires processing a multitude of inputs at once, like identifying objects in a complex scene. Analog circuits have a natural flair for parallel processing, akin to having a team of experts tackling different aspects of the problem simultaneously. Digital systems, while incredibly efficient for specific tasks, may require additional steps or hardware to achieve the same level of parallelism.

And here's where adaptability comes into play. Think of a dynamic environment where you need to continually adapt to changing conditions. Analog hardware can mimic this dynamic process, allowing for continuous adaptation and learning. It is like being able to refine your skills as you go, seamlessly adjusting to new information. Digital systems can certainly learn, but they might require a bit more scaffolding to match the fluidity of analog hardware.

Let us make a list of technologies and concepts used in neuromorphic technologies. Future neuromorphic devices will likely be hybrid between one or more of these technologies and CMOS-based architectures:

1. **Memristors**: Memristors are resistors with memory. They retain a memory of past resistance states, which makes them a key component in mimicking synaptic behavior in neuromorphic systems.

2. **Phase-Change Memory (PCM)**: PCM technology utilizes the reversible phase change of a material between amorphous and crystalline states. This can be employed to emulate synaptic weights in neural networks.

3. **Neuromorphic Processors**: Dedicated neuromorphic processors, like IBM's TrueNorth, are designed to efficiently execute neural network computations. They are optimized for tasks such as pattern recognition and sensory processing.

4. **Optoelectronic Neuromorphic Systems**: These systems combine optical and electronic elements to enable high-speed processing. Optical components can mimic the transmission of signals in biological neurons.

5. **Neuromorphic Sensors**: These specialized sensors are designed to mimic the behavior of sensory receptors in biological organisms. They play a crucial role in enabling neuromorphic systems to interact with their environment.

6. **Analog-Digital Hybrid Architectures**: Hybrid architectures combine analog and digital components, leveraging the strengths of both paradigms. This approach seeks to balance efficiency and precision in neuromorphic computing.

7. **Nano-scale Neuromorphic Devices**: Nanoscale devices, such as nanowires and nanotubes, are explored for their potential to emulate neural behavior due to their unique electronic properties.

8. **Biohybrid Neuromorphic Systems**: These systems integrate biological components, such as living neurons or neuromorphic cells, with artificial hardware to create hybrid computational platforms.

9. **Optical Computing**: Optical computing architectures are investigated for their potential in emulating neural processing. Optical elements can enable parallel and energy-efficient computations.

10. **Synthetic Neurobiology**: This interdisciplinary field combines neuroscience and synthetic biology to create engineered neural circuits and components for use in neuromorphic systems.

We had already discussed memristors and SNNs. Let us however discuss in depth the other technologies.

PCM technology leverages the unique property of certain materials to transition between amorphous (disordered) and crystalline (ordered) states. This phase change is reversible, allowing the material to store binary data. In the amorphous state, the material exhibits high electrical resistance, representing a '0'. When crystallized, it has low resistance, representing a '1'. By precisely controlling the transition, PCM can encode and retain information. This property makes PCM an attractive candidate for emulating synaptic weights in neural networks, mostly because of the low power transition property. The ability to store and update these weights with high precision and endurance aligns with the requirements of neuromorphic systems. PCM's non-volatility (meaning it retains data even when power is removed) and fast switching times further enhance its suitability for applications in artificial intelligence and computational neuroscience.

Neuromorphic processors are specialized computing units designed to mimic the behavior of biological neural networks. We have already discussed Loihi and SpiNNaker. Unlike conventional central processing units (CPUs) or graphics processing units (GPUs), which follow a digital von Neumann architecture, neuromorphic processors are optimized for tasks inspired by the brain's computational principles. These processors comprise arrays of simple, low-power processing elements that can perform parallel computations. Each element is designed to emulate a neuron or a small group of neurons, complete with synaptic connections. What sets neuromorphic processors apart is their ability to perform in-situ computation, meaning they process information directly where it is stored, much like neurons in biological systems. One of the most notable examples of a neuromorphic processor is IBM's TrueNorth. It features a massive array of spiking neurons and synapses, enabling it to efficiently process information in a manner akin to the brain's neural networks. TrueNorth excels at tasks like pattern recognition, sensory processing, and other applications that demand real-time, low-power computation. These processors represent a departure from the traditional digital von Neumann architecture. They are optimized for tasks that benefit from massive parallelism and low energy consumption, making them well-suited for applications in artificial intelligence, robotics, and various domains where real-time processing and low power consumption are paramount.

Optoelectronic neuromorphic systems represent an innovative approach to computing that combines principles of optics and electronics. These systems leverage the advantages of both domains to achieve high-speed and energy-efficient information processing. In optoelectronic neuromorphic systems, optical elements are integrated with electronic components to perform computations. This integration allows for the manipulation of information using photons, enabling parallel processing and high-speed data transmission. Optical components, such as lasers, waveguides, and photodetectors, play a crucial role in facilitating communication and computations. One of the key advantages of optoelectronic systems lies in their ability to transmit information at the speed of light. This can lead to significantly faster processing times compared to purely electronic systems. Additionally, the parallelism inherent in optics allows for the simultaneous manipulation of multiple data streams, which is particularly advantageous for tasks that require high throughput. Optoelectronic neuromorphic systems hold promise for a range of applications, including artificial intelligence, signal processing, and high-performance computing. They are especially well-suited for tasks that demand rapid decision-making or the processing of large volumes of data in real-time.

While still an evolving field, ongoing research and development in opto-electronic neuromorphic systems aim to harness the full potential of this interdisciplinary approach to computing. By seamlessly integrating optical and electronic components, these systems are poised to play a pivotal role in shaping the future of computational technology.

Neuromorphic sensors also represent a cutting-edge class of sensors designed to emulate the behavior of sensory receptors in biological organisms. These sensors are engineered to capture and process information from the environment in a manner akin to how biological organisms perceive the world. At their core, neuromorphic sensors integrate specialized hardware that mimics the principles of biological sensory systems. They are equipped with sensor elements that respond to various stimuli, such as light, sound, pressure, or temperature. These sensors then transduce these stimuli into electrical signals, similar to how sensory organs in living organisms convert external stimuli into neural impulses. What sets neuromorphic sensors apart is their ability to process information directly at the sensor level. This in-situ computation capability allows them to perform initial processing steps on the raw sensory data, akin to how sensory neurons in biological organisms preprocess incoming stimuli. This feature is particularly valuable in scenarios where real-time processing and low-latency responses are critical. For example, in a neuromorphic vision sensor, pixels are equipped with processing elements that can detect specific features or patterns directly within the sensor itself. This enables tasks like edge detection or motion tracking to be executed at the sensor level, reducing the computational load on downstream processing units. Neuromorphic sensors find applications in fields ranging from computer vision and robotics to environmental monitoring and healthcare. Their ability to efficiently process sensory information at the sensor level holds great promise for tasks that demand rapid and contextually relevant responses. As this field advances, ongoing research is focused on refining the design and capabilities of neuromorphic sensors. By leveraging principles from both neuroscience and sensor technology, these sensors are poised to play a pivotal role in enabling machines to perceive and interact with their environments in a manner that closely parallels biological organisms.

Analog-Digital hybrid architectures represent another cutting-edge approach to neuromorphic computing that combines the strengths of both analog and digital processing elements. These architectures aim to achieve a synergistic balance between the precision of digital computation and the efficiency and adaptability of analog circuits. In analog-digital hybrid architectures, the system is designed to leverage the unique advantages of both analog and digital components. Analog processing elements excel at tasks that involve continuous, real-valued representations and operations. They can process signals in a smooth, gradient-like manner, akin to how neurons communicate through graded potentials. This makes analog circuits well-suited for tasks that require nuanced, real-valued computations, such as sensory processing or feature extraction. On the other hand, digital components are adept at managing discrete, well-defined states and executing Boolean logic operations with high accuracy. They are robust against noise and can maintain consistent computational precision over time. Digital circuits are particularly valuable for tasks that involve complex control flows, decision-making, and tasks that demand high computational accuracy. The key advantage of analog-digital hybrid architectures lies in their ability to harness the complementary strengths of both domains. By strategically integrating analog and digital elements, these architectures can achieve superior performance and energy efficiency compared to using either paradigm in isolation. For example, in a neuromorphic system, analog components may be employed for front-end processing tasks, such as sensory data preprocessing or feature extraction. These analog circuits can efficiently process raw sensory information and extract relevant features. Subsequently, the digital components can take over for higher-level decision-making and control tasks, where discrete states and precise computations are crucial.

The integration of analog and digital elements in hybrid architectures opens up new possibilities for efficient and versatile neuromorphic computing, and should be in fact the future of computing. As researchers continue to explore and refine this approach, analog-digital hybrid architectures hold great promise for enabling machines to perform complex tasks with a level of efficiency and adaptability that was once challenging to achieve using either analog or digital computation alone.

Let us now discuss other nanoscale devices, particularly nanowires. Nanowires have diameters on the order of nanometers and long of approximately a few micrometers, and exhibit unique electrical properties that make them exceptionally promising for a wide range of applications, including neuromorphic computing. Nanowires are one-dimensional structures composed of semiconductor materials like silicon, germanium, or compound semiconductors such as gallium arsenide. The latest generation of nanowires are made of a coated conductor, such as silver. Their nanoscale dimensions introduce quantum mechanical effects that significantly influence their electronic behavior. These effects become particularly pronounced as the size of the nanowire approaches the scale of the electron's wavelength. One of the key advantages of nanowires is their high surface-to-volume ratio. This feature enhances their sensitivity to surface interactions, making them well-suited for sensing applications. In the context of neuromorphic computing, nanowires hold significant potential for emulating the synaptic connections between neurons. By functionalizing the surface of nanowires with appropriate materials, researchers can create memristive devices. These devices exhibit dynamic resistance changes in response to voltage pulses, mimicking the behavior of biological synapses. We will go more into the details of these devices in the next chapter. Their small size and high integration density enable the construction of densely interconnected networks that operate on principles similar to biological neurons. Moreover, the compatibility of nanowires with complementary metal-oxide-semiconductor (CMOS) technology facilitates their integration into existing electronic platforms, allowing for seamless incorporation into larger-scale systems.

Despite their immense potential, challenges remain, including precise control over nanowire synthesis, reproducibility, and scalability. Additionally, ensuring uniform electrical properties across a large number of nanowires is a critical consideration for practical applications. As researchers continue to refine fabrication techniques and explore novel material combinations, nanowires are poised to play a crucial role in the development of advanced neuromorphic computing systems.

Biohybrid neuromorphic systems represent a cutting-edge approach to computing that merges biological components with artificial electronics, creating a synergistic platform that combines the strengths of both natural and engineered systems. At the core of biohybrid systems are biological elements, such as living neurons, neuronal networks, or even entire brain slices, which interact with synthetic electronic components. These biological components bring the unparalleled information-processing capabilities of biological systems to the forefront. They are capable of performing complex computations and learning tasks through mechanisms like synaptic plasticity. Complementing the biological components are synthetic elements, typically comprised of electronic circuits or neuromorphic devices. These components provide the necessary interface between the biological and artificial domains, facilitating communication, control, and data processing. They may include memristors, transistors, or specialized neuromorphic hardware. One of the key advantages of biohybrid neuromorphic systems lies in their ability to harness the adaptability and learning capabilities of biological neurons. By interfacing living neurons with synthetic electronic elements, researchers can create platforms capable of biological-like learning and adaptive behavior. This opens up exciting opportunities for the development of advanced neuromorphic computing platforms. Biohybrid systems have found applications in a range of fields, including neurobiology research, neuroprosthetics, and brain-computer interfaces. They enable scientists to study the behavior of biological neurons in controlled environments, offering insights into fundamental neurophysiological processes. However, challenges remain, including the need for robust and stable interfaces between biological and synthetic components, as well as the ethical considerations surrounding the use of living organisms in research and technology development. As research in biohybrid neuromorphic systems advances, the potential for creating intelligent systems that seamlessly integrate biological and synthetic elements holds great promise. These systems have the potential to revolutionize fields such as neurobiology, brain-machine interfaces, and neuromorphic computing, paving the way for innovative applications that bridge the gap between biology and technology.

Optical computing represents another approach to information processing that utilizes photons, or light particles, as carriers of information,
in contrast to traditional electronic computing which relies on electrons.
This paradigm leverages the unique properties of light, such as its speed,
parallelism, and coherence, to perform computations. At the heart of optical computing are optical elements, which include components like lasers,
lenses, mirrors, and modulators. These elements manipulate the properties of light to perform various computational tasks. For instance, lasers
generate coherent light beams, while modulators control the intensity or
phase of light waves. One of the most promising applications of optical
computing is in neuromorphic systems. Optical components can emulate
the behavior of neurons and synapses, facilitating the creation of optical neural networks. These networks process information in a manner
that closely mimics the parallelism and distributed processing seen in biological neural networks. Optical computing excels at tasks that involve
massive parallelism, such as matrix operations and Fourier transformations. This makes it particularly well-suited for applications like image
processing, cryptography, and complex simulations. Optical computing
also holds significant potential for overcoming the limitations of traditional electronic computing, such as speed and energy efficiency. Photons travel at the speed of light, enabling extremely high data transfer
rates. Additionally, optical systems can perform computations in a highly
parallel manner, potentially accelerating certain tasks by orders of magni-
tude. However, there are challenges to be addressed in the development of
optical computing systems. These include issues related to light sources,
the integration of optical components with electronic devices, and the
need for robust optical interconnects. Despite these challenges, advances
in materials science and photonics technology continue to drive progress
in optical computing. Researchers are exploring innovative approaches,
such as photonic integrated circuits and metasurfaces, to enhance the capabilities of optical computing further. As optical computing continues
to evolve, it holds the promise of revolutionizing the way we process and
manipulate information. Its potential impact spans a wide range of fields,

from high-performance computing to artificial intelligence, opening up new frontiers in computational capabilities.

Synthetic neurobiology represents a cutting-edge interdisciplinary field that merges principles from neuroscience, biology, and engineering to create artificial systems that mimic and interact with the nervous system. This field seeks to unravel the mysteries of neural processing while also engineering novel solutions for interfacing with and augmenting biological systems. One of the key goals of synthetic neurobiology is to understand the fundamental principles that govern neural function. By creating synthetic models of neurons and neural circuits, researchers can probe the intricate dynamics underlying information processing in the nervous system. These models range from simplified electronic circuits that mimic neurons to more sophisticated simulations that replicate entire neural networks. Another crucial aspect of synthetic neurobiology involves the development of neural interfaces. These interfaces serve as bridges between artificial devices and the nervous system, enabling bidirectional communication. They can take the form of implantable electrodes, microfluidic systems, or even optogenetic tools that use light to control neural activity. Furthermore, synthetic neurobiology plays a pivotal role in advancing neuroprosthetics and brain-machine interfaces (BMIs). By leveraging synthetic components, researchers can create devices that restore lost sensory or motor functions in individuals with neurological disorders or injuries. These prosthetic systems can decode neural signals, interpret intentions, and translate them into meaningful actions, such as moving a robotic limb.

In addition, synthetic neurobiology has opened up exciting possibilities for studying and manipulating brain function. By introducing synthetic elements into the brain, researchers can modulate neural activity, study neural plasticity, and even induce specific patterns of neural firing. This has implications for both basic research in neuroscience and potential therapeutic interventions. Despite its immense potential, synthetic neurobiology poses ethical and safety considerations, particularly when interfacing artificial components with the human nervous system. Ensuring the biocompatibility, long-term stability, and safety of neural interfaces remains a critical area of research. As the field of synthetic neurobiology continues to advance, it holds great promise for revolutionizing our understanding of the brain and developing transformative technologies that could enhance human health and capabilities. By blending the realms of biology and engineering, synthetic neurobiology stands at the forefront of interdisciplinary research, with profound implications for both neuroscience and neurotechnology.

Architectures for neuromorphic devices play a crucial role in harnessing the capabilities of emerging technologies. One prominent architecture that has garnered significant attention is the Crossbar Array.

The Crossbar Array architecture represents a grid-like arrangement of intersecting wires, where each intersection point forms a memristive device. Memristors are two-terminal passive devices that exhibit a non-volatile resistance change in response to applied voltage. They serve as crucial elements for emulating synaptic connections in neuromorphic systems.

In a Crossbar Array, the vertical and horizontal wires represent the inputs and outputs, respectively. When a voltage is applied across a particular pair of wires (forming a crosspoint), it modulates the resistance state of the memristive device at that crosspoint. This alteration in resistance mimics the strengthening or weakening of a synaptic connection in biological neural networks.

One of the key advantages of Crossbar Arrays lies in their high density of synaptic connections. With each crosspoint representing a synapse, these architectures can achieve extremely high integration levels. This density is essential for emulating the vast number of synapses present in biological brains.

Furthermore, Crossbar Arrays exhibit inherent parallelism in their operation. Multiple synaptic connections can be modified simultaneously, mirroring the parallel processing capabilities of biological neural networks. This feature is crucial for achieving real-time, energy-efficient computations in neuromorphic systems.

However, there are challenges associated with Crossbar Arrays, including issues related to sneak-path currents, variability in memristor characteristics, and the need for precise control of voltage levels. Researchers are actively addressing these challenges to improve the reliability and performance of Crossbar Arrays.

In addition to Crossbar Arrays, other architectures, such as neuromorphic processors and hybrid analog-digital systems, are also being explored. These architectures combine various types of neuromorphic devices and computational elements to achieve specific computational goals.

As research in neuromorphic computing progresses, architectures like Crossbar Arrays continue to play a pivotal role in the development of high-performance and energy-efficient neuromorphic systems. Their potential applications span a wide range of fields, including artificial intelligence, pattern recognition, and brain-machine interfaces, offering a promising pathway towards advanced neuromorphic computing platforms.

The advantage of Crossbar Arrays for memory addressability stems from their dense and highly interconnected structure. In a Crossbar Array, each intersection point represents a potential synaptic connection, allowing for an enormous number of individual connections to be formed within a relatively compact space. This high density of synapses enables Crossbar Arrays to address a vast amount of memory locations simultaneously.

Moreover, due to the potential for three-dimensional of crossbar arrays, with multiple layers of crosspoints stacked on top of each other, they can achieve an unprecedented level of volume efficiency. This means that a significant amount of synaptic connections can be stored in a relatively small physical space. As a result, Crossbar Arrays offer an unparalleled capacity for memory storage and processing, making them highly desirable for applications that demand large-scale neural network simulations and computations.

The combination of dense interconnectivity and volume efficiency in Crossbar Arrays provides a powerful platform for emulating the complex synaptic connections found in biological neural networks. This attribute is crucial for tasks like pattern recognition, machine learning, and other computational tasks that benefit from the parallel and distributed processing capabilities of neuromorphic systems. Additionally, the volume efficiency of Crossbar Arrays contributes to their energy efficiency, as it reduces the distances over which signals must be transmitted, minimizing power consumption.

Going back to In-Memory computing, one of the key advantages of In-Memory Computing is its potential to dramatically reduce data movement and alleviate the "memory wall" problem that traditional computing architectures face. By processing data in-place, In-Memory Computing can lead to significant gains in terms of speed and energy efficiency.

Moreover, In-Memory Computing is particularly well-suited for tasks that involve massive parallelism and data-intensive operations, such as pattern recognition, machine learning, and certain types of simulations. It has found applications in domains like artificial intelligence, data analytics, and scientific computing.

One notable example of In-Memory Computing is the use of resistive switching devices, such as memristors, which can perform both memory storage and computation within the same device. This enables a seamless integration of memory and processing, paving the way for more efficient computing architectures.

While In-Memory Computing offers tremendous promise, it's important to note that its implementation presents technical challenges, including issues related to device reliability, integration complexity, and programming models. Nevertheless, ongoing research and advancements in materials science and device technology continue to drive progress in this exciting area of computing.

While In-Memory Computing represents a paradigm shift from the traditional von Neumann architecture by performing computations directly within the memory, crossbar arrays fall short of this idea. In the next chapter we will discuss alternative options and possible directions to overcome the simplicity of crossbar architectures.

6 The future of neuromorphics

While Crossbar Arrays offer significant advantages for certain applications, they may face challenges when employed as the backbone of in-memory computing architectures.

One prominent concern is the issue of scalability. As the size of Crossbar Arrays increases, so does the complexity of controlling and addressing individual memristive devices. This can lead to challenges in terms of signal-to-noise ratios, power consumption, and the ability to reliably read and write data. In practice, scaling Crossbar Arrays to extremely large sizes may be non-trivial.

Another limitation arises from the sneak-path problem. In large Crossbar Arrays, unintended current paths can form due to the highly interconnected nature of the structure. These sneak paths can lead to crosstalk and interference, potentially compromising the integrity of stored data.

Furthermore, variability in memristive devices is a critical issue. Each memristor may have slightly different characteristics, such as resistance levels and switching behavior. This variability can affect the accuracy and reliability of computations performed within the array, particularly in precision-sensitive applications.

In addition, endurance and reliability are important considerations. Memristive devices have finite lifetimes, with a limited number of read-/write cycles before degradation occurs. This can be a significant concern in applications that require frequent and sustained operations.

Finally, there are challenges related to programming and control. Achieving precise and reliable programming of memristive devices, especially in large-scale arrays, can be a non-trivial task. Developing efficient algorithms and control schemes that account for device variability and non-idealities is an active area of research.

One of the fundamental distinctions between Crossbar Arrays and the architecture of the brain lies in their underlying principles. Crossbar Arrays rely on resistive switching elements, such as memristors, to store and process information. These elements operate based on electrical resistance and are typically binary, representing a stark contrast to the complex, analog nature of synapses and neurons in the brain.

In the brain, synapses exhibit a vast range of analog strengths, allowing for nuanced and graded information processing. This analog nature enables the brain to perform tasks such as pattern recognition, learning, and adaptation with remarkable efficiency. Crossbar Arrays, on the other hand, often operate in a binary mode, lacking the continuous range of synaptic strengths found in biological systems.

Moreover, the brain's architecture is highly distributed and massively parallel, with billions of neurons interconnected through an intricate network of synapses. This enables the brain to process information in a highly parallelized and fault-tolerant manner. In contrast, Crossbar Arrays are constrained by their grid-like structure, which may limit their ability to replicate the decentralized, massively parallel processing characteristic of the brain.

Another critical distinction is the brain's ability to exhibit plasticity and adaptability through mechanisms like long-term potentiation (LTP) and long-term depression (LTD). These processes allow the brain to modify the strength of synaptic connections in response to learning and experience. While there have been efforts to emulate plasticity in memristive devices, achieving the level of sophistication and adaptability seen in biological systems remains a significant challenge.

Furthermore, the brain exhibits a high degree of fault tolerance and resilience to component failures. In contrast, Crossbar Arrays may be more susceptible to reliability issues, especially as they scale up in size and complexity.

In contrast to Crossbar Arrays, memristive systems incorporating self-organizing nanowires exhibit characteristics that align more closely with the architecture of the brain.

Firstly, self-organizing nanowires have the potential to emulate certain aspects of the brain's structural plasticity. In the brain, synaptic connections can dynamically form, strengthen, or weaken in response to learning and experience. Self-organizing nanowires can mimic this process by autonomously growing and forming new connections in response to electrical stimuli or environmental cues. This ability to adapt and reconfigure connectivity shares a fundamental principle with the brain's capacity for synaptic plasticity.

Furthermore, the analog nature of resistive switching in self-organizing nanowires provides a closer approximation to the continuous range of synaptic strengths observed in biological synapses. This analog behavior allows for graded information processing, which is crucial for tasks like pattern recognition and associative memory, mirroring the brain's computational abilities more effectively than binary systems.

Additionally, the distributed and decentralized nature of self-organizing nanowires can resemble the massively parallel architecture of the brain. Unlike traditional grid-like structures, self-organizing systems can exhibit a high degree of connectivity and redundancy, similar to the intricate neural network of the brain. This decentralized organization enhances fault tolerance and adaptability, aligning more closely with the robustness of biological systems.

Moreover, the potential for self-organizing nanowires to undergo morphological and structural changes in response to external stimuli mirrors the brain's ability to undergo synaptic remodeling. This capacity for structural adaptation contributes to the system's resilience and adaptability, further paralleling the brain's architecture.

In brain-like architectures, learning is a intricate process that involves a dynamic interplay between inputs, outputs, and the system's inherent plasticity. This process is fundamentally distinct from simply relying on internal memories.

One key aspect is that learning in brain-like architectures is highly influenced by external stimuli and interactions with the environment. The system's plasticity, which allows for the modification of synaptic strengths, is primarily driven by the patterns and characteristics of incoming sensory inputs and responses to external cues. This dynamic response to external stimuli enables the system to adapt and acquire new information, a crucial characteristic of learning.

Additionally, the brain exhibits a form of unsupervised learning, where it can identify underlying patterns and structures in data without explicit labels or guidance. This ability to discover meaningful features from raw sensory information is essential for tasks like perception and pattern recognition. In contrast, relying solely on internal memories may not provide the same level of adaptability and contextual understanding.

Furthermore, learning in brain-like architectures is often associative, allowing the system to form connections between different elements based on their co-occurrence or proximity in time. This enables the system to establish relationships between various inputs and outputs, facilitating tasks like memory retrieval and pattern completion.

It's important to note that while internal memories play a role in the learning process, they serve as a repository of prior experiences rather than the primary driver of learning. Internal memories serve as a contextual backdrop against which new information is processed and integrated, allowing for a richer and more nuanced understanding of incoming stimuli.

As of the present, there is a noticeable lack of emphasis on theoretical analysis and the development of toy models in the field of neuromorphic computing. Several factors contribute to this observation.

Firstly, the rapid pace of technological advancement in neuromorphic hardware has led to a predominant focus on empirical experimentation. With the emergence of increasingly sophisticated neuromorphic devices, researchers are understandably eager to explore their capabilities and potential applications. This emphasis on empirical validation often takes precedence over theoretical groundwork.

Furthermore, the inherent complexity of neuromorphic systems presents a challenge for analytical modeling. Unlike conventional digital computing, which is based on well-established theoretical frameworks, neuromorphic architectures involve intricate interactions between various elements, including memristive devices, synaptic connections, and neural units. The nonlinear and dynamic nature of these systems can make it difficult to formulate analytically tractable models.

In addition, the field of neuromorphic computing is inherently inter-disciplinary, drawing expertise from areas such as materials science, electrical engineering, neuroscience, and computer science. This multidisciplinary nature has led to a prevailing inclination towards experimental and applied research, with a primary focus on building functional hardware prototypes and exploring real-world applications.

Moreover, the empirical approach aligns with the field's objective of developing neuromorphic systems that can perform specific tasks, such as pattern recognition or sensory processing. By directly implementing and testing these systems, researchers can gauge their effectiveness in practical applications.

While this emphasis on empirical validation is crucial for advancing the field and demonstrating the capabilities of neuromorphic hardware, there is indeed a need for a complementary focus on theoretical analysis and toy models. Theoretical frameworks and simplified models can provide valuable insights into the fundamental principles governing neuromorphic behavior. They can serve as conceptual tools for exploring the capabilities and limitations of different architectures, as well as for gaining a deeper understanding of the underlying computational mechanisms.

Nanowire systems pose unique challenges when it comes to the application of Kirchhoff's laws, which govern the flow of electrical currents in circuits. These challenges arise due to the nanoscale dimensions and unique physical properties of nanowires.

Firstly, at the nanoscale, quantum effects become significant. Unlike in macroscopic circuits, where electrons can be treated as classical particles, in nanowires, quantum mechanical phenomena such as tunneling and ballistic transport can dominate electron behavior. This necessitates the use of advanced quantum transport models to accurately describe electron flow, which go beyond the classical Kirchhoff's laws.

Furthermore, the surface-to-volume ratio in nanowires is much higher compared to bulk materials. This leads to an increased influence of surface states, which can introduce additional complexities in current flow. The behavior of electrons near the surface may deviate from bulk behavior, requiring a careful consideration of boundary conditions and surface effects.

Additionally, nanowires often exhibit non-uniform cross-sectional geometries and may possess heterostructures with varying material properties along their length. This spatial variation in properties can introduce local variations in resistances and capacitances, making it challenging to apply Kirchhoff's laws uniformly.

Moreover, in certain nanowire systems, the presence of defects, impurities, or dopants can introduce scattering mechanisms that affect electron transport. These scattering events may not be adequately captured by classical circuit models, necessitating the incorporation of quantum mechanical scattering theory.

Finally, the small size of nanowires can lead to increased thermal effects, such as self-heating. These thermal effects can influence the electrical behavior of the nanowire and may require the consideration of coupled thermal-electrical models, which go beyond the scope of traditional Kirchhoff's laws.

The integration of a neuromorphic processing unit alongside traditional CPUs represents a transformative approach to computing that holds tremendous potential across a spectrum of applications.

Firstly, neuromorphic processing units excel in tasks that demand parallel processing, pattern recognition, and learning. By harnessing the principles of artificial neural networks, these units can rapidly process complex data sets in real time. This capability makes them highly suited for tasks ranging from image and speech recognition to natural language processing and autonomous navigation.

Furthermore, neuromorphic processing units are inherently energy-efficient. Their operation is inspired by the low-power, parallel processing observed in the human brain. By leveraging memristive and synaptic devices, these units can perform computations with significantly lower energy consumption compared to traditional von Neumann architectures. This makes them ideal candidates for applications in edge computing, IoT devices, and scenarios where power efficiency is paramount.

Additionally, the adaptability and learning capabilities of neuromorphic units render them versatile across a broad range of tasks. Through training and reconfiguration, these units can dynamically adapt to new tasks and environments without the need for extensive reprogramming or retooling. This level of adaptability is particularly advantageous in scenarios where tasks are not predefined or may evolve over time.

Moreover, integrating a neuromorphic processing unit alongside CPUs offers the potential for hybrid computing architectures. This enables tasks to be offloaded to the neuromorphic unit, where they can be executed in parallel, freeing up the CPU for other computational tasks. This can lead to significant gains in overall system performance and efficiency.

In applications such as gaming, where real-time responsiveness and complex pattern recognition are critical, a neuromorphic processing unit can provide a distinct advantage. By training the unit specifically for gaming tasks, it can rapidly process sensory input, adapt to dynamic game environments, and enhance overall gameplay experiences. This is only one simple example, but in much more important situations neuromorphic computing might become relevant.

While integrating neuromorphic processing units holds great promise, it is imperative to recognize that realizing the full potential of these architectures necessitates the concurrent development of novel algorithms specifically designed to leverage their unique capabilities.

Firstly, traditional algorithms, which are primarily designed for von Neumann architectures, may not fully exploit the parallelism and adaptability inherent in brain-like systems. To unlock the true power of neuromorphic units, algorithms must be tailored to harness the distributed, massively parallel nature of neural networks. This requires a departure from conventional algorithmic paradigms and the exploration of new computational frameworks inspired by the brain.

Moreover, brain-inspired architectures often operate on principles of spiking neural networks, which differ significantly from the continuous-valued computations of traditional digital computing. Algorithms need to be reimagined to accommodate the discrete, event-driven nature of spiking neurons. This includes the development of spike-based encoding, decoding, and learning mechanisms that can effectively process information in the temporal domain.

Additionally, the plasticity and adaptability of brain-like systems necessitate the creation of learning algorithms that can dynamically adjust synaptic strengths in response to input patterns. These algorithms should be capable of unsupervised, reinforcement, and supervised learning, mirroring the diverse learning mechanisms observed in biological systems.

Furthermore, as brain-like architectures often incorporate memristive and neuromorphic hardware, algorithms must be designed to interface seamlessly with these specialized devices. This includes accounting for the unique characteristics and constraints of memristive elements, such as resistance-switching dynamics and non-linear conductance behaviors.

In applications such as pattern recognition, natural language processing, and autonomous decision-making, developing algorithms tailored for brain-like architectures becomes crucial. These algorithms must be capable of harnessing the system's adaptability, fault tolerance, and energy efficiency to deliver superior performance compared to traditional computational approaches.

Neuromorphic computing, with its intricate interplay of hardware, software, and neural-inspired principles, demands a paradigm of understanding and development akin to a complex adaptive system rather than a fully engineered one.

First and foremost, the behavior of neuromorphic systems emerges from the collective interactions of a multitude of components, each with its own dynamic response. This emergent behavior cannot be fully predicted or prescribed by a purely deterministic engineering approach. Instead, it requires an understanding of how local interactions give rise to global phenomena, akin to the behavior of complex adaptive systems in nature.

Moreover, neuromorphic systems possess a high degree of adaptability and self-organization. They are capable of learning, reconfiguring, and adapting to changing environments and tasks, mirroring the dynamic nature of biological neural networks. Attempting to engineer every aspect in a top-down, rigid manner would stifle this inherent adaptability, hindering the system's ability to learn and evolve in response to its surroundings.

Furthermore, the complexity and richness of neuromorphic behavior cannot be fully captured through traditional engineering models alone. The integration of memristive devices, spiking neural networks, and specialized hardware components introduces non-linearities, stochasticity, and emergent phenomena that are best understood and harnessed through approaches rooted in complexity science.

In addition, the development of algorithms for neuromorphic systems often relies on principles of self-organization, unsupervised learning, and reinforcement learning. These algorithms enable the system to learn from its environment and make autonomous decisions, mirroring the adaptive behavior observed in complex natural systems. Attempting to engineer every aspect of the learning process from scratch would stifle the system's ability to adapt and thrive in dynamic environments.

Let us offer an alternative point of view. Neuromorphics should be the field dedicated to emulating the architecture and functionality of biological neurons and synapses in artificial systems. The focus should lie on a system that employs a network of 'learning devices' able to replicate the behavior of neurons and synapses found in the human brain. For instance, nanowires are minuscule, measuring about one-thousandth the width of a human hair, and are typically composed of a highly conductive metal, such as silver, coated with an insulating material like plastic.

These nanowires have the remarkable ability to self-assemble into a network structure that bears resemblance to a biological neural network. Nanowires autonomously organize themselves, creating a network architecture reminiscent of a biological neural network. Much like neurons, which possess insulating membranes, every metal nanowire is sheathed in a slender insulating layer

Upon applying electrical signals to these nanowires, ions migrate across the insulating layer and into adjacent nanowires, akin to the transmission of neurotransmitters across synapses. This leads to the emergence of synapse-like electrical signaling within the nanowire networks.

Recent results from researchers from the University of Sydney show that learning in this way of possible. However, the deep connection between the field of cybernetics and this new line of research is missing.

Concepts like homeostasis, control, the role of complexity and adaptation, are yet concepts to be explored.

To conclude, the integration of neuromorphic processing units and the development of algorithms tailored for brain-like architectures herald a new era in computing. With their potential for parallelism, adaptability, and energy efficiency, these systems hold immense promise for revolutionizing a wide array of applications, from artificial intelligence to edge computing and beyond. As researchers continue to push the boundaries of both hardware and software, the future shines bright with the possibilities that brain-inspired computing brings to the forefront of technological advancement.